BIOLOGICAL
RESOURCE
CENTERS

BIOLOGICAL RESOURCE CENTERS

Knowledge Hubs for the Life Sciences

Scott Stern

BROOKINGS INSTITUTION PRESS
Washington, D.C.

Copyright © 2004
THE BROOKINGS INSTITUTION
1775 Massachusetts Avenue, N.W., Washington, D.C. 20036
www.brookings.edu

Library of Congress Cataloging-in-Publication data

Stern, Scott, 1969–
    Biological resource centers : knowledge hubs for the life
sciences / Scott Stern. — 1st ed.
        p.   cm.
    Includes bibliographical references (p.) and index.
    ISBN 0-8157-8149-0 (pbk. : alk. paper)
    1. Biological resource centers. I. Title.
    QH324.15.S74   2004
    570'.72—dc22   2004021431

9 8 7 6 5 4 3 2 1

The paper used in this publication meets minimum requirements of the
American National Standard for Information Sciences—Permanence of Paper
for Printed Library Materials: ANSI Z39.48-1992.

Typeset in Adobe Garamond

Composition by Betsy Kulamer
Washington, D.C.

Printed by R. R. Donnelley
Harrisonburg, Virginia

Contents

Acknowledgments

I am particularly indebted to the many scientists who graciously offered their insights, the personnel of the American Type Culture Collection, and especially Raymond Cypess and Robert Hay.

Bob Litan and Michael Kremer were instrumental in helping this project come to fruition and provided helpful guidance throughout, and the volume could not have been completed without the help of my collaborator Jeff Furman. I would also like to thank Tim Bresnahan, Joshua Gans, Shane Greenstein, Rebecca Henderson, Paul Romer, Cathy Fazio, an anonymous referee, and participants in numerous seminars and workshops. Jason Corradini, Mercedes Delgado, Lorraine DeLeon, Chijoke Emineke, Anna Harrington, Martha Kam, Julia Lo, Kinga Piekos, Megan MacGarvie, Courtney Mason, Tracy Myers, and Naomi Wohl provided excellent research assistance and Catherine Theohary checked the volume for accuracy.

Finally, thanks are extended to the Brookings Institution Press and its staff in shepherding the book through to publication, in particular Jim Schneider, Eileen Hughes, and Janet Walker for editing the manuscript. Some of this research I conducted as a visiting fellow at the Brookings Institution, whose hospitality is gratefully acknowledged. All errors of course remain my own.

Biological Resource Centers

1

Introduction

This book examines the history and explores the impact of biological resource centers, the world's "living libraries." These biological resource centers collect, store, and distribute varied biological materials—including cell lines, microorganisms, recombinant DNA material, media, and reagents—and provide the information technology tools that allow researchers to access the biological materials. Because these centers authenticate, preserve, and offer access to the materials that underlie new knowledge, they allow results produced by one generation of researchers to be validated and built on by follow-on researchers. Biological research centers (BRCs) enhance the research process itself by providing a transparent and standardized way of accessing biological materials, facilitating step-by-step discovery in the life sciences.

The growth and impact of BRCs as knowledge hubs for the life sciences can be understood only by appreciating the importance of the fidelity, or integrity, of biological materials in life sciences research. Consider the so-called "HeLa scandals."[1] Before researchers grasped the importance of fidelity, most biomaterials were routinely exchanged between research labs through a peer-to-peer network. Though most researchers assumed that the

1. Gold (1986); Masters (2002).

1

exchanges simply offered a low-cost means of building on discoveries made by others, the system resulted in what many consider the most far-reaching methodological failure in modern life sciences research. Without specific methods for certifying the authenticity of biomaterials, a surprisingly large share of those exchanged between labs became contaminated with other cell lines or otherwise misidentified. A scientist using materials from the laboratory of another researcher who might have believed that he or she was conducting an experiment on a cell line derived from healthy embryonic lung cells of a white male might, in fact, have unwittingly been using cancerous cervix cells from a thirty-one-year-old black woman. That is precisely what happened. In a series of revelations that came to be known as the HeLa scandals (named after a particularly robust cervical cancer cell line responsible for a large number of contaminations), certification errors cast doubt over thousands of individual research findings worldwide, including the research of Nobel Prize–winning scientists.[2]

This was not simply an isolated mistake; contamination and misidentification of biological materials still plague published (and patented) research findings.[3] To take but one example, the KB cell line, originally derived from oral cancer cells, was contaminated and taken over by HeLa. It is well documented that researchers using the KB line are actually performing experiments with the HeLa cervical cancer cell line. More than 300 articles based on research using the KB cell line were published between 1998 and 2000, many of them claiming to provide new findings relating to oral cancer.[4] Moreover, the findings of many of these articles have themselves been extensively cited by subsequent researchers in support of their own research. The persistence of misidentification arises at least in part because there are powerful incentives for researchers conducting scientific and commercial research to claim priority for a novel discovery or finding, but there are few if any incentives to validate research claims made by others.

Though contamination poses a risk to the integrity of all microbiology research, only a relatively small group of scientists, many of them historically involved with or employed by BRCs, have addressed the problem in a comprehensive way. Because of efforts to develop systematic tests for cell culture identification and to establish certified standard cell culture reference collections at these centers (tasks that are far from complete), microbiologists have been able to limit their exposure to contamination. By helping to identify and rectify methodological failures, restore trust in the fidelity of biomateri-

2. Gold (1986); Masters (2002).
3. Masters and others (2001).
4. Masters (2002).

als, and establish new standards for access to authenticated research materials, BRCs have come to occupy a central place within the life sciences research infrastructure.[5]

Yet biological resource centers remain largely invisible and neglected by policymakers; their function as knowledge hubs promoting discovery and innovation in the life sciences based on cumulative knowledge is neither widely recognized nor well understood. This book provides the first systematic economic assessment of the impact of these centers. BRCs offer a case study of the importance of and requirements for step-by-step scientific and technological progress and of the impact of institutions on cumulative knowledge production. Through a detailed examination of the BRCs' position within the life sciences infrastructure, this book identifies a new policy agenda for all stakeholders in the life sciences community.

Cumulative Knowledge Production: The Source of Long-Term Economic Growth

Cumulative scientific progress and technological innovation are central to long-run gains in productivity and economic achievement. More than any single invention or discovery, continuous innovation holds the key to maintaining prosperity and ensuring a healthy and stable rate of economic growth. Over time, technology-driven growth has increasingly become centered on sectors and industries with strong reliance on science and particular scientific disciplines. The dramatic improvements over the last half century in computers and telecommunications, for example, are rooted in the physical sciences; even today, improvements in technologies ranging from semiconductors to cell phones rely on specific advances in underlying scientific disciplines, from physics to chemistry to materials science.

The impact of science-driven cumulative innovation is perhaps most apparent in industries and technologies with connections to the life sciences. Although biology and related disciplines had some impact before World War II, the last sixty years have witnessed an amazing expansion in both the scope of the life sciences and their application to pressing social and human problems. And the revolution is far from over. The sequencing of the human genome and the extraordinary research opportunities opened up by stem cells are but heralds of an era of continuing fundamental discovery and promise.

This promise has attracted rapidly growing public and private investment. In the past quarter century, life sciences research has come to account for

5. OECD (2001); Cypess (2003d).

more than 50 percent of total federal research expenditures and an increasing overall share of U.S. research and development expenditures. Relative to most other sectors that have increasingly taken on an applied R&D focus, the life sciences continue to maintain a high level of investment in basic research, conducted within a network of university laboratories, government institutions, and private companies.

In large part the progress that occurs within this network has its roots in the norms and methods of science. At the crux of the scientific method is the idea that discoveries by one researcher can be verified and built upon by later researchers. As Isaac Newton said, when scientific norms operate effectively, researchers continually "stand on the shoulders of giants," in pursuing their own discoveries. For example, James Watson and Francis Crick's discovery of DNA's double helix structure in 1953 depended on previous research (such as the X-ray crystallography experiments of Rosalind Franklin) and laid the foundations for the continuing revolution in molecular biology.

Despite its centrality, the role of specific institutions and public policy in encouraging cumulative knowledge production has received little attention. The conditions that foster the accumulation of knowledge are subtle. The mere production of a piece of knowledge in no way guarantees that others will be able to exploit it. To make progress, researchers must both be aware of earlier findings and believe in their truthfulness. Uncertainty or lack of awareness of the findings necessitates costly reverification and reinterpretation, reducing the productivity of current R&D efforts. To be effective, cumulative innovation must somehow enable researchers to draw, rely, and build on an ever larger body of scientific and technical knowledge.

The Role of Knowledge Hubs in Cumulative Knowledge Production

Centralized institutions—here called knowledge hubs—are of decisive importance in facilitating cumulative research, and biological resource centers are primary knowledge hubs for the life sciences. A cumulative knowledge hub is an institution or system enabling knowledge transfer among researchers through its centralized deposit and access facilities. Knowledge hubs perform three central functions: acquisition, authentication, and access. Knowledge hubs provide facilities for researchers to deposit the materials necessary to verify research results, thus eliminating the need for future involvement of the original researcher. They then certify the materials, through replication or other authentication mechanisms. Finally, they provide access to the materials to future researchers. Without effective knowledge hubs, the process of accumulating knowledge begins to stagnate and returns on investment in R&D begin to diminish.

Public policy performs a crucial but often overlooked role in the development and effectiveness of knowledge hubs. More than the outcome of a specific research grant, effective policy is essential for ensuring continued progress in a particular field; it allows expenditures and attention to be leveraged, leading to improvements in the quality of a knowledge hub that increase the productivity of all members of a research network. Science and innovation policy must be premised on a clear understanding of the role that knowledge hubs play in research and the policy mechanisms that encourage their sustained growth and effectiveness.

Because BRCs promote cumulative knowledge in life sciences research, effective policy toward them is crucial to maintain productivity and lower research costs.

Plan of the Book

To illustrate the role of biological resource centers in the life sciences, chapter 2 outlines their origin, evolution, and adaptation over the past century. The chapter highlights the exchange of biomaterials, the challenges in ensuring the integrity of the research process, and the role of collective institutions such as BRCs in addressing those challenges. While different BRCs vary in their operations, they are united in seeking to improve the research process by providing access to authenticated biomaterials. Since effective experimental design depends on understanding detailed properties of research materials (to rule out alternative effects or mechanisms), scientific progress can be enhanced by access to materials whose properties have been characterized by others whose work can be trusted. Understanding how institutions affect confidence in research materials and the implications for overall research productivity requires assessing the role of institutions in cumulative progress.

Chapter 3 addresses this challenge by highlighting the impact of BRCs on scientific research in their roles as knowledge hubs. The centers perform four essential functions in this regard. First, by certifying and authenticating materials deposits, they enhance the validity of research findings. Second, they contribute to cumulative knowledge by providing long-term preservation and storage of biomaterials. By maintaining this stock, BRCs allow materials associated with discoveries that may have little initial value to be accessed effectively when complementary technology or developments allow their full value to be realized. Third, transactions in a peer-to-peer network often suffer from incentive and contracting problems, but BRCs allow for low-cost access by a wide audience of later researchers. Fourth, the centers possess economies of scale and scope in collecting and distributing materials. These functions are at the heart of what makes BRCs function so effectively as knowledge hubs for the life sciences, enhancing the cumulative capacity of the life sciences research community.

Chapter 4 empirically investigates the impact of BRCs on the diffusion and use of scientific knowledge, presenting a citation analysis for articles linked to BRC deposits. The leading biological resource center in the United States, the American Type Culture Collection (ATCC), collects bibliographical reference information for the majority of deposits in its collection. To the extent that depositing biological materials in BRCs increases the accessibility of that knowledge for future research, then scientific articles associated with BRC deposits should be more intensively cited as a result of their greater impact on follow-on research. The results of this empirical analysis are dramatic. BRC-linked articles have a 200 percent higher likelihood than others of being cited, a gap that becomes larger with the length of time since initial publication. A second stage of the analysis isolates the marginal impact of BRCs by showing that deposit in a BRC resulted in a doubling of future citations, mostly as a result of additional citations more than five years after deposit. This boost from deposit in a BRC became more pronounced during the 1990s. Finally, the chapter assesses how expenditures on BRC deposits compare with alternative research investments in promoting cumulative progress. While the average "cost" per citation for university-based biological research is estimated at $2,400, the cost per citation from deposit in a BRC is less than $900. Relative to traditional grant activities and considered at current levels of expenditures, investments in BRC authentication, storage, and access activities are nearly threefold more efficient in promoting step-by-step cumulative advances in biology and the life sciences.

Chapter 5 provides a systematic analysis of policy implications for government agencies that support life sciences research, the scientific and commercial research communities, and BRCs themselves. For each group, the knowledge hubs framework implies a principle providing overarching policy guidance. Each principle can then be applied to generate detailed policy recommendations for the functional activities performed by BRCs.

First, the *cumulative knowledge investment principle* states that enabling the cumulative process should be a priority for government intervention in life sciences research. Project evaluation should consider not only the quality of research proposals, but also the terms on which the results and resources developed in the projects are made available to the scientific community. Among other consequences, this principle suggests that research funding should be far more proactive in ensuring the accessibility of research resources and that depositing materials in a BRC should be treated as de facto compliance with new federal grant requirements. Similarly, federal investment in strengthening the integrity of research resources should be substantially expanded.

Second, the *cumulative knowledge participation principle* emphasizes the responsibility that must be borne by the academic and commercial research communities in ensuring the strength and vitality of BRCs. These responsibilities include providing access to the widest possible range of authenticated biomaterials as well as establishing research rules and procedures and making investments that enable the accumulation of knowledge within the life sciences. For instance, to increase access to materials that are protected by intellectual property laws, university and industry associations should promote the use of material transfer agreements for international patent materials deposited in BRCs. The research community must police itself by adopting procedures and reward systems that encourage early detection of published research errors and appropriate authentication of biomaterials.

Third, the *cumulative knowledge performance principle* highlights the importance of high standards, operational efficiency, and reputation to BRCs themselves. Enhancements to BRCs improve research productivity and lower research costs. This principle supports standardized authentication protocols through an international BRC network. Established BRCs should provide assistance in creating new BRCs, especially in developing countries.

Chapter 6 addresses some policy implications for the entire life sciences community. The analysis suggests that both the public and private sectors must substantially increase their investment and participation in ensuring the effectiveness of the biomaterials exchange system. A strong life sciences research infrastructure depends on sustained investment. These observations have implications for a range of contemporary policy issues facing the scientific community, including the role of bioinformatics, the rise of biotechnology, the potential for biodiversity efforts, and the new challenges of bioterrorism.

Overall, this book provides an economic assessment of the role played by biological resource centers in life sciences research. Along with related institutions promoting cumulative knowledge production, BRCs serve as knowledge hubs for life sciences research. While this role makes BRCs critical to the life sciences research enterprise, public awareness about these institutions remains low. However, lack of attention by the broader public should not be confused with potential policy impact. In particular, policymakers have relatively few levers available to accelerate the rate of scientific and technical progress. Focusing on those institutions that enable researchers to "stand on the shoulders of giants" is a particularly promising approach.

2

Evolution and Adaptation:
A Short History of
Biological Resource Centers

The life sciences revolution is still in its infancy. But compared with other scientific and technological pursuits, the life sciences have grown dramatically over the past few decades. Most media coverage focuses on particular scientific discoveries or on the ethical and economic concerns raised by such breakthroughs. However, the impact of scientific and commercial research depends crucially on the presence of institutions to facilitate the growth and use of knowledge over time. Despite the lack of headlines, these institutions—and the policies that nurture their growth and development—are essential to maintaining an effective research system.

The role of biological resource centers in supporting scientific progress is more than an abstract concern. Indeed, their impact cannot be understood without appreciating a practical but central challenge for life sciences research: maintaining the integrity of shared biomaterials. Without an effective system for ensuring that materials used in experiments are the materials that researchers *believe* they are using, microbiological research could not advance. But this deceptively simple problem has bedeviled the research community since the advent of transportable biomaterials in the 1950s. While the most well-known cases of misidentification were uncovered during the so-called HeLa scandals of the 1970s, contamination and misidentification of biological materials remains a central problem (box 2-1). According to recent estimates, perhaps more than 20 percent of all cell lines remain misidentified, and thousands of articles based on misidentified cell lines are published every year.[1] The history of BRCs is inextricably intertwined with these authentication problems. This chapter offers a brief glimpse into that fascinating history in order to begin to identify the impact of BRCs on the research process itself.

It is important to emphasize that maintaining the fidelity of research materials is not principally a technological or scientific problem; instead, it is a problem of incentives. Scientists have few incentives to engage in replication and validation of others' research, but individual scientists may have strong incentives to ensure their own financial advantage and professional prestige by thwarting the process of validating (or not) their own previous research. Thus although maintaining a robust system for validating experimental research is in the interest of all scientists, individual researchers may have little stake in contributing to this outcome. The integrity of the scientific process is a public good, and in large part BRCs are an institutional response designed to overcome this conflict between public goods and private rewards.[2] The remainder of this chapter explores biomedical research centers as economic institutions, first by surveying the functions of biological resource centers, then tracing their evolution and adaptation over time and highlighting the challenges they face.

What Is a Biological Resource Center?

A BRC is an institution dedicated to the investigation of the structure, functions, and applications of biological systems by developing cooperative activities for sharing validated biological material and data among government agencies, industry, academia, and the public.[3] As a key element of the life sciences research infrastructure, the centers maintain large and varied collections of biological materials—including cell lines, microorganisms, recombinant DNA material, media, and reagents—and provide the information technology tools that allow researchers to access these materials. During the past quarter century, BRCs have come to play an increasingly important role in scientific and commercial research. For example, since the 1980s, certain centers, such as the American Type Culture Collection (ATCC), have been critical in protecting intellectual property rights by serving as international patent depositories for all patented living organisms.

1. MacLeod and others (1999); Masters (2002).
2. Jong (2003).
3. Cypess (2003a).

The concept and operation of BRCs evolved from the development of culture collections, also referred to as cell banks. Robert Stevenson, a former director of the ATCC, defined culture collections as officially constituted organizations "acquiring, preserving, authenticating, and distributing microorganisms or in vitro cultured cells to qualified scientists."[4]

Since the early 1980s, the level of investment in life sciences research has skyrocketed, and increasing importance has been placed on the availability of reliable and standardized biomaterials. This resulted in pressure to expand the functions of traditional culture collections to include systematic standardization and authentication of research materials, the development of research tools to enhance the productivity of research using standardized materials, and proactive knowledge management for both public and private entities. More than culture collections of repositories, the BRC concept includes the role of these institutions as "complex scientific organizations which acquire, authenticate, preserve, produce, develop and share biological materials, information, technology, intellectual property, and standards for the advancement, validation, and application of scientific knowledge."[5]

Though the concept of a BRC was initially controversial, the importance of expanding the scope of culture collections is now widely acknowledged. For example, in a recent report from the Organization for Economic Cooperation and Development (OECD), the broadened mandate of BRCs includes their role as "service providers and repositories of living cells, genomes of organisms, and information relating to heredity and the function of biological systems. BRCs contain collections of culturable organisms (for example, microorganisms, plant, animal, and human cells), replicable parts of these (for example, genomes, plasmids, viruses, cDNAs), viable but not yet culturable organisms, cells and tissues, as well as databases containing molecular, physiological and structural information relevant to these collections and related bioinformatics."[6] An ongoing debate continues over the extent of transformation required by culture collections before they may be considered full-fledged BRCs.

In part, debate about the precise scope of BRCs reflects marked differences in their functioning and governance. As of 2004, more than 480 BRCs were registered with the global BRC federation, the World Federation of Culture Collections. Of these, approximately one-third are in Europe, and another third are in Asia. Less than one-quarter of BRCs registered with the

---

4. Stevenson and Hatt (1992, vol. 1, p. 15).

5. Cypess (2003b, p. x).

6. Organization for Economic Cooperation and Development (2001).

Box 2-1. A Conspiracy of Cells

In 1951 Henrietta Lacks, a young black Baltimore woman, died of cervical cancer. A tissue sample of her cancer lived on, however, revolutionizing microbiology and simultaneously giving rise to a devastating research contamination scandal. Shortly after Lacks's cancer diagnosis, the husband and wife team of George and Margaret Gey, at Johns Hopkins University, obtained a sample of her cancerous tissue, code named HeLa. Before HeLa, scientists had struggled to get human tumor cells to live in vitro. HeLa, however, was different. Incubated in a feeding solution in the Gey lab, the HeLa cells thrived, and they proved their durability even after being shipped through the U.S. mail. Biologists and other life science researchers enthusiastically adopted the HeLa line, and there was rapid growth in the number of "HeLagrams" being sent among laboratories in the United States and abroad. By being the first and most durable "tumor in a test tube," HeLa was quickly established as the single most popular human tissue culture in the world.

The success of HeLa inspired scientists to culture other tissue samples, and soon researchers had a range of different types of human tissue cell lines to work with. Throughout the 1950s and early 1960s, cell culture science flourished. Unfortunately, with so many different types of cell lines circulating worldwide, an occasional mistake was unavoidable. Sometimes tissue cultures were mixed through mislabeling, but more often through cross-contamination. If two distinct cell lines ended up in the same petri dish (perhaps as the result of unintentional transfer by a researcher working with both cell lines in the same lab), the stronger of the two lines would take over, and the researcher would be none the wiser because proper screening was impossible due to the difficulty in characterizing cells.

By the early 1960s the contamination and identification problem was well recognized. In 1962 the National Cancer Institute sponsored an initiative to house the nation's best set of reference cells at the ATCC. These cells had to be of the highest quality and their history clearly documented. Nevertheless, doubts continued to linger, because there was still no definitive way to identify cell lines. In 1966, however, geneticist Stanley Gartler developed the first robust method for consistent identification of tissue culture.

Gartler's research focused on certain proteins and enzymes. He realized that some enzymes varied predictably in every human being. In his results, however, he was surprised to discover that many of the established cell lines he tested had identical enzyme patterns. One enzyme was particularly telling since it occurred only in black people, but most cell lines were reportedly from Caucasian patients. Gartler realized that the HeLa cell line had contaminated a substantial part of all laboratory-based cultures in the United States (and potentially abroad), with the result that much of the research done by these labs was either invalid or must at least be reconfirmed. At that time a relatively unknown geneticist, Gartler produced research that demonstrated that respected biologists from leading institutions had unwittingly conducted years of invalid research. Perhaps not surprisingly, this conclusion was met with some skepticism by those involved; it was not until two independent teams confirmed the findings in 1968 (by showing that twenty-four of thirty-six "distinct" ATCC human cell lines were in fact all HeLa) that biology researchers took the contamination threat seriously.

After initial panic and some precautionary measures (more careful identification and control, development of tests, and so forth), most researchers treated the contamination problem as a thing of the past. While caution and care in experimentation had to be intensified, science could move forward. However, things were not so simple. In 1973 Walter Nelson-Rees, the head of the Cell Culture Laboratory at Berkeley, used chromosonal markers to revisit the question of cell misidentification, and his lab discovered that the contamination problem was both widespread and deeply rooted. In 1974 Nelson-Rees published the first of several "hit lists" in *Science*, identifying research findings and laboratories that he had found were likely to have been using contaminated or misidentified tissue cultures. By and large the culprit was HeLa; even at the height of the cold war, with little contact between U.S. and Soviet bloc scientists, HeLa cells were found to contaminate Russian labs (causing, among other things, a minor diplomatic incident). Although press reports declaring a major setback in the "War on Cancer" from HeLa cell contamination may have been overstated at times, the work of Nelson-Rees and others demonstrated that even the best laboratories could easily mishandle biological materials. Lewis Coriell, for example, demonstrated that merely pulling a stopper from a test tube or dispensing liquid from a dropper could launch tiny airborne HeLa droplets that had the potential to overtake other in vitro cell cultures. Even polio vaccine inventor Jonas Salk's lab was contaminated for several years. The cost of the damage to cumulative research in the life sciences from these mistakes has not been and cannot be truly estimated. Each report of misidentification reduced the confidence of researchers exchanging materials, forcing ever greater replication of results and a corresponding decline in research productivity.

The title of this box and this short discussion borrow heavily form Michael Gold's *A Conspiracy of Cells*, which describes the misidentification and cross-contamination phenomenon as well as the crucial role played by contamination detectives such as Walter Nelson-Rees and Robert Stevenson (Gold, 1986).

Table 2-1. *Types of Biological Resource Centers*

Center type	Examples
Public nonprofit national collections — Tier 1	ATCC (USA) DSMZ (Germany) Japan Collection of Microorganisms (Japan)
Public nonprofit specialized collections — Tier 2	Coriell Medical Research Institute (human genetic mutant cell lines) National Institute of Allergy and Infectious Disease (HIV materials) Ribosomal Database Project Agricultural Research Service Culture Collection (NRRL)
Private industrial collections — Tier 2	Merck (antibiotics screening collection, clinical microbiology collection) Institute for Fermentation Osaka (IFO)
Specialized university collections — Tier 2	Yale University (*Escherichia coli* Genetic Stock Center) Ohio State (*Bacillus* Genetic Stock Center) Penn State (*Fusarium* Genetic Stock Center)
Life sciences data management institutions	Ribosomal Database Project (RDP) Institute for Genomic Research (IGR)

Source: Adapted from Organization for Economic Cooperation and Development (2001, pp. 25–27).

World Federation of Culture Collections are in the Americas.[7] Collections vary dramatically in their size, ranging from fewer than 100 to more than 100,000 cultures, resulting in a hierarchy of BRC types (table 2-1). Overall, more than 1 million cultures are maintained by centers worldwide, and millions of cultures are distributed each year to scientists throughout the world. While some countries maintain very small collections numbering in the hundreds, the biological resource centers in the United States maintain nearly 200,000 cultures.[8] Although large national collections such as the ATCC in the United States and the DSMZ in Germany manage a broad range of materials, specialized collections, such as Yale University's *E. coli* collection, focus on providing a niche resource for specific scientific areas. But although all collect, identify, and distribute materials, BRCs vary in how exhaustively they authenticate and certify the materials in their collections. For example,

7. See the World Federation for Culture Collections, Culture Collections in the World (wdcm.nig.ac.jp/statistics.html [August 2004]).
8. See note 7.

Figure 2-1. *Source of Funding for Culture Collections*

Number of collections

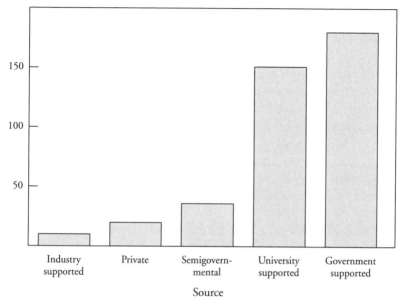

Source: Author's calculations based on information gathered at http://wdcm.nig.ac.jp/hpcc.html (May 2004).

while leading BRCs in the United States such as the ATCC and the Coriell Institute are recognized for their international leadership in cell culture technique, newer and smaller collections have tended to adopt techniques and protocols developed elsewhere.[9] Finally, while a plurality of BRCs around the world are supported mainly by governments, the large national collections in the United States rely on a mix of public and private support (figure 2-1). For example, the ATCC, organized as a private not-for-profit institution, receives less than 15 percent of its total budget through direct government grants; the majority of its income is derived from services such as patent depository and distribution services.

Despite differences in size and funding, BRCs have a common purpose. They serve as intermediaries for knowledge transfer, making materials and research results developed by one generation of researchers available to future researchers. The centers enhance the validity of research itself by providing a transparent and standardized procedure for accessing biological materials.

9. Biological resource centers in the former Soviet Union, for instance, have been beset with a series of funding and maintenance crises, exemplified by the near shutdown of the Obolensk national collection in early 2002. See Tyler (2002).

The value created by these functions arises from the way biological research is conducted. Biological research depends on the development and conduct of experiments that allow researchers to disentangle alternative hypotheses about the composition and functioning of living organisms. In many cases effective experimental design depends on understanding detailed properties of a biological organism in order to rule out alternative effects and mechanisms. By using biological materials whose properties have been characterized by previous researchers and that can be accessed through a BRC, scientists may reduce experimental uncertainty—the uncertainty associated with the scientific tests themselves.

As economic institutions BRCs may increase research and development productivity by providing independent access to a wide variety of standardized biological materials. Although the entire research community benefits from this enhanced environment, individual researchers have few incentives to increase the productivity of other researchers. As a result, the benefits of the centers may not be realized in the absence of interventions or collective action to support their core activities.

The Evolution and Growth of BRCs

Although cell cultures have been used in baking and in fermentation for thousands of years, the value of a central repository for such materials became apparent only with the rise of bacteriology in the late nineteenth century. In 1890 at the Institute of Hygiene at the Faculty of Medicine in Prague, František Král established what may be considered the first culture collection that could be accessed by other researchers.[10] Král's Bakteriologisches Laboratorium, which supplied cultures of bacteria, yeasts, filamentous fungi, and media to laboratories and schools, published its first catalogue in 1902.

The increase of international culture collections was rapid, largely in response to the enormous increase in the number of biological researchers and laboratories throughout the world. The Centraalbureau voor Schimmelcultures of the Netherlands, founded in 1904, published its first catalogue in 1907.[11] Japan was not far behind, establishing its first collection in 1904 with the creation of the Japanese Society for Culture Collections. The first British collection, the National Collection of Type Culture, was established in 1920 at London's Lister Institute.[12]

10. Stevenson (1992).
11. Centraalbureau voor Schimmelcultures (www.cbs.knaw.nl/about/index.htm [May 2004]).
12. Stevenson (1992).

In the United States the Society of American Bacteriologists (now the American Society for Microbiology) was responsible for the first materials collections as early as 1899. C. E. A. Winslow, a prominent society member, proposed a "museum of living bacteria for the benefit of working laboratories all over the world."[13] This vision was realized in 1911 with the establishment of a collection at the Museum of Natural History in New York. While BRCs have changed in scope and substance since their origin as "living museums," several features of their earliest history remain relevant: the collections were established through deposits received from public and private collections from around the world, both deposit and access were offered at the lowest possible cost, and the collections could not survive without the constant attention of committed individuals. The Natural History Museum collection was prosperous for a time, with 578 donations by 1912, but the collection deteriorated from lack of supervision after Winslow moved to Yale University. By 1922, only 175 strains were viable.[14]

The Natural History Museum collection was the direct predecessor of the American Type Culture Collection, which was established by a consortium of scientific societies in 1925. The ATCC, recognizing a need for a centralized depository in the United States, sought "to preserve cultures of microorganisms that have historic and scientific interest and to provide a center for obtaining cultures needed in education and research."[15] Initially housed at the McCormick Institute in Chicago, the ATCC collection grew rapidly in its early years, despite facing severe financial constraints with the onset of the Great Depression.[16] Growing to over 2,000 strains, the ATCC was the premier culture collection in the United States by the time it moved to the Georgetown University School of Medicine in 1937.

The early, entrepreneurial years of biological resource centers were marked by reliance on certain individuals to nurture collections and by the absence of a model for systematic financial support. However, BRCs, like other scientific institutions, adapted and matured in response to the massive expansion in attention to and support of scientific research during and after World War II. From 1947 onward the ATCC, as the premier U.S. culture collection, was formally incorporated as a nonprofit organization and has consistently maintained its own facilities in the greater Washington, D.C., area. In part this evolution tempered reliance on specific individuals (for example, most BRCs

13. Quoted in Clark and Geary (1974, pp. 295 and 297).
14. Clark and Geary (1974).
15. Hay (2003).
16. For example, ATCC staff were paid on a month-to-month basis, based on current revenues (Clark and Geary, 1974).

elected a board of directors to manage transition issues) and the frequency of funding crises (centers began charging private companies for access to materials in order to cover recurring overhead expenses). At the same time, these changes ushered in enduring challenges for BRCs. Three stand out: the extraordinary and sustained growth of life sciences research; the increasing use of transportable biomaterials and establishing the integrity of biomaterials; and the establishment of different funding regimes for life sciences research infrastructures throughout the world.

Together, these challenges have shaped the development of BRCs as life sciences institutions.

The Growth of the Modern Life Sciences

There has been perhaps no greater single influence on BRCs than the multiple transformations and steady growth of the life sciences over the past half century. Spurred by innovations such as penicillin, both the federal government and private researchers dramatically increased their investment in research in the life sciences during and after World War II. From the founding of the National Science Foundation, the National Institutes of Health, and other agencies after World War II, the central importance of institutions supporting scientific research and innovation has been recognized and incorporated into federal policy.[17] Since then, the United States has been able to maintain international leadership in the life sciences, in no small part because of the increasing public and private commitment to supporting those institutions (figure 2-2).

For their part, BRCs have actively extended the scope and nature of their activities in response to advances in biology. For example, although the biochemical origins of mutation were not identified until the early 1940s, BRCs were quick to recognize that this phenomenon called for the systematic development and maintenance of stocks of different genetic strains of organisms.[18] Similarly, with the rise of the study of bacteriophages in the late 1940s and early 1950s, the ATCC received a grant from the newly formed National Science Foundation to establish a bacteriophage collection, which has grown dramatically and remains a core collection of these viruses. In 1959, with NIH funding, the ATCC established the Viral and Rickettsial Registry, which facilitated the accessibility of virus stocks for researchers throughout the United States.[19]

17. Nelson (1993); Nelson and Rosenberg (1994).
18. Stevenson (1992).
19. Hay (2003, pp. 155–56).

Figure 2-2. *Federal R&D Funding, by Budget Function, 1980–2001*

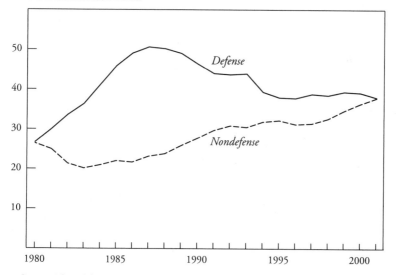

Billions of constant 1996 dollars

Billions of constant 1996 dollars

Source: Adapted from *NSF Science and Engineering Indicators, 2002.*

a. "Other" includes all nondefense functions not separately graphed, such as agriculture and transportation. The 1998 increase in general science and decrease in energy resulted from a reclassification.

Beyond simply responding to scientific advances, BRCs established standardized and accessible materials, facilitating more rapid scientific and medical advancement. Consider the identification of the gene for Huntington's chorea. In 1979 Nancy Wexler, a leading Huntington's researcher, began the systematic collection of tissue samples from the residents of a Venezuelan town with a high incidence of the disease. These cells, deposited and stored at the Coriell Institute in New Jersey, were distributed to researchers around the world and were crucial in identifying the location of the gene for the disease in 1993.[20] In other words, the ability of researchers to push the limits of knowledge depends on their ability to access biological materials crucial to emerging research. By establishing themselves as essential to this process, BRCs grew dramatically in size, scope, and importance during the rapid expansion of the life sciences that began in the 1940s.

The Advent of Transportable Biomaterials and Maintaining the Integrity of Biomaterials

The development of BRCs is inextricably linked to the increasing use of transportable biomaterials and to the need to establish the integrity of biomaterials as well as to the overall growth in the life sciences. Two technologies led to those developments: in vitro cell culture and storage techniques. Indeed, it was not until the early 1950s, with the establishment of the HeLa cell line by the Gey Laboratory at Johns Hopkins University, that biologists could continuously grow and maintain human tissue culture outside the body. Moreover, the development of freeze-dry techniques in the 1940s and advances in the 1960s in vapor phase and liquid nitrogen technology allowed materials to be stored indefinitely.[21] These advances allowed for a dramatic expansion in the range of materials that could be transported between laboratories worldwide.[22] Overall, these changes offered dramatic opportunities for systematic scientific progress as researchers increased the power of their experiments by exploiting findings of previous research using identical biological material. Most notably, in certain subdisciplines a few microorganism types became established as model systems by the late 1950s. In the development of bacteriology, for example, a small number of bacteriophage strains

20. "Institute Overview," Coriell Institute (http://arginine.umdnj.edu/about/cimrsumm. html [May 2004]).

21. It is useful to note that BRC researchers were instrumental in these developments. For example, liquid nitrogen storage technology was pioneered at the ATCC through the research of Shuh-Wei Hwang and others (Clark and Geary, 1974).

22. Clark and Geary (1974); Hay (2003).

whose genetic composition was well understood by researchers were exploited in most key experiments. However, in vitro culture and storage technology also created a crisis: the inability to consistently identify or avoid contamination of cultures imported from other laboratories. Despite recognition of the problems in the late 1950s by some scientists, they were consistently underestimated by most life sciences researchers during the 1960s and 1970s. Most dramatically, laboratory cultures in the United States (and throughout the world) were contaminated by the HeLa cell line, resulting in the invalidation of years of publicly funded biological research, some by leading scientists. Though they were more attentive to the problem than all but a few university-based scientists, even prominent BRCs initially fell victim to the misidentification scandals: among thirty-four cell lines at the ATCC as of 1968, twenty-four turned out to be replicants of HeLa.[23]

Thus, although advances in the technology underlying biological research increased scientists' ability to draw on knowledge and materials developed by previous researchers, mechanisms for ensuring the authenticity of materials were initially inadequate. Eventually, BRCs became key institutional players in establishing tools and procedures for overcoming these obstacles. As early as 1964 the ATCC moved to a state-of-the-art facility to establish itself as a source for credible standard reference materials.[24] And Walter Nelson-Rees at the Cell Culture Laboratory at the University of California, Berkeley, spearheaded efforts over a decade to eradicate misidentification and cross-contamination from published studies.[25] Perhaps more than any other action, BRCs' efforts to address the problem of misidentification and to restore trust in biomaterials exchange established the importance of their role in life sciences research.

The misidentification challenges continue. On one hand, under the leadership of directors Robert Stevenson and Raymond Cypess since the early 1980s, the ATCC has undertaken a number of initiatives, including the development of tests and markers for culture identification, to establish reliable and certified biological materials to laboratories throughout the United States.[26] And the DSMZ in Germany has established a program to ferret out the use of falsely labeled cell lines in research, including by systematically testing materials used in highly cited scientific research articles. Unfortu-

23. See box 2-1; Gold (1986).
24. The success of these early efforts is ambiguous. Gold suggests that at least some leading researchers, including Stevenson, believed that centers did not live up to their potential in avoiding misidentification and contamination during the 1970s (Gold, 1986, p. 135).
25. Gold (1986).
26. Hay (2003).

nately, these investments have not guaranteed the integrity of biomaterials. The DSMZ estimates that approximately 20 percent of all cell lines used in tumor research are misidentified, and literally thousands of studies based on faulty cell lines have been published.[27] While there is considerable debate among scientists about the impact of misidentification today, it remains disturbing that at least some articles using falsely identified materials are themselves cited very often. The consequences of misidentification can throw doubt on multiple generations of research findings.

The Emergence of Alternative BRC Funding Regimes

BRCs' efforts to support life sciences research have been hampered by uneven and uncertain funding. Some BRCs have always faced severe funding shortfalls, sometimes resulting in the discontinuation of certain collections (often with a chilling effect on research in that area). In response, various centers and national governments have experimented with alternative funding sources in order to identify a viable financial support model. For example, by establishing themselves as standard reference collections and undertaking certification activities, BRCs obtain most of their materials without paying the depositor. This arrangement has allowed centers both in the United States and abroad to maintain a greater variety of materials at a lower cost of acquisition.

Despite their broad similarity, the funding models for BRCs vary dramatically across collections and have changed over time. Large national collections in the United States rely principally on patent depository and culture distribution fees; national collections in Europe rely much more heavily on direct government grants. Less than 10 percent of the ATCC's current budget is from direct government grants, but nearly 100 percent of the budget of Germany's DSMZ comes from government sources.[28] Many smaller collections are supported by specific donors or very narrow grant programs; for example, the Rhizobium Research Laboratory maintains its collection in part with funds from the Minnesota Department of Transportation.[29]

The trend away from direct government subsidies in the United States is the consequence first of wide swings in National Institutes of Health (NIH) funding in the 1960s and early 1970s (there was a 50 percent reduction in the research appropriation for NIH between 1966 and 1967) that encouraged centers to expand the share of their income derived from activities that

27. MacLeod and others (1999); Masters (2002).
28. Organization for Economic Cooperation and Development (2001).
29. See Sponsors List, Rhizobium Research Laboratory (www.rhizobium.umn.edu/sponsors.htm [May 2004]).

Figure 2-3. *National Center for Research Resources: Percentage of Federal Life Sciences Research Funds, 1938–2000*

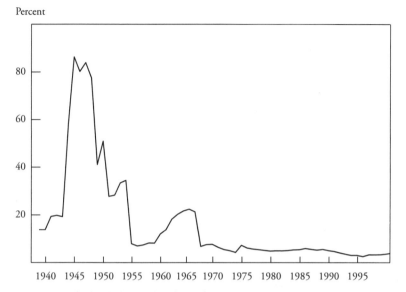

Percent

Source: Author's calculations, based on information gathered at www.nih.gov/about/almanac/appropriations/index.htm.

did not directly depend on such support (figure 2-3). As a result, during the 1980s the ATCC dramatically improved its cataloguing and distribution capabilities, experiencing an increase of more than 700 percent in cell line distribution activities.[30] Second, the Budapest Treaty made the deposit of biological materials a requirement for obtaining a patent as of 1980. Through patent depository fees, centers serving as international patent depositories have been able to increase their financial viability and the range of services they provide. Finally, BRCs have increasingly served as repositories for private collections of pharmaceutical, biotechnology, and chemical companies. By charging for the use of their advanced storage equipment, they have added a potential source of continuing support.

Although BRCs around the world are funded through a variety of mechanisms, leading U.S. centers have combined multiple sources to create a funding regime that depends only modestly on direct government subsidy. Of course, as new challenges emerge, the rationale for this arrangement should be revisited. Given the potential importance of BRCs in improving

30. Hay (2003).

scientific productivity, are funding policies that impinge on their financial health wise?

Emerging Challenges

Four primary challenges confront the further evolution of BRCs. First, because microbiology and genomics have converged, effective research requires access to both biological materials and sophisticated databases and software. Second, since biotechnology has come of age, traditional pharmaceutical companies and new biotechnology firms are creating and using cutting-edge biological materials in their commercial product development efforts. Third, the increased threat of bioweapons development by terrorists and others has dramatically shifted the debate about the control and distribution of sensitive bioorganisms. Fourth, efforts to assign and preserve ownership rights over the full range of naturally occurring biological materials have given rise to new requirements (as well as new opportunities) for control of and access to biomaterials (as well as to biodiversity concerns). Considering each of these challenges is crucial in assessing the effectiveness of policies to promote BRCs' continued growth.

Bioinformatics: Information Technology in the Life Sciences

Information technology is transforming almost every sector in the global economy, and life sciences research is no exception. The combination of massive storage capacity, powerful data manipulation techniques, and graphics capabilities has revolutionized both how research is conducted and how the resulting knowledge is preserved. The very structure of genetic information—extremely long sequences of a code made up of four letters—is ideally suited to information technology (IT). The emerging discipline of bioinformatics is in fact defined as one deriving "knowledge from computer analysis of biological data . . . [including] information stored in the genetic code, but also experimental results from various sources."[31] Effective exploitation of the IT revolution, however, requires careful integration of electronic and physical data resources.

Just as the ability to sustain cultures in vitro was crucial to the development of fields such as virology, information technology and the potential of bioinformatics are transforming traditional fields such as molecular biology and spawning new fields such as genomics and proteomics.[32] Since the initial

31. Michael Nilges and Jens P. Linge, "Bioinformatics—A Definition" (www.pasteur.fr/recherche/unites/Binfs/definition/bioinformatics_definition.pdf [May 2002]).
32. Freshney (2000).

establishment of the GenBank database in 1982, for example, the amount of genetic and biological information stored and available for retrieval in electronic databases has exploded. As of the end of 2003, there were more than 30 million sequences stored at GenBank, identifying more than 36 billion DNA base pairs.[33] This growth is the result, in part, of the development and diffusion of research tools such as polymerase chain reaction (PCR) and sequence-tagged sites in the 1980s (developments made more timely by BRC collections) and by the expansion in funding for sequencing for the Human Genome Project.

But while the sequencing of the human genome is the most visible accomplishment of genomics, the scientific and clinical value of databases such as Genbank and the Human Genome Project depends crucially on the ability to integrate insights from electronic databases with a detailed understanding of biomaterials. In the absence of certified biological materials to conduct genomic research, the value of sequencing is ambiguous at best. At the same time, the availability of advanced tools for identifying and interpreting gene sequences increases the return on maintaining reference cell lines. For example, identifying the Huntington's chorea gene required both access to very specialized cell cultures and advanced information technology tools to analyze the genomic differences in them. However, integrating biological and electronic data has been difficult. Despite widespread public acclaim following the definitive sequencing of the human genome in April 2001, researchers are still identifying large-scale errors requiring reevaluation and correction.[34] Moreover, establishing a precise match between biological and electronic data sources is not simply a technological problem; until the agreement between NIH and Celera Genomics to cooperate on completing the human genome in early 2001, scientists from each group routinely accused the other of sacrificing scientific precision for the sake of speed.[35]

From the earliest days of the genomics revolution, BRCs have been at the forefront in establishing the databases to link electronic and physical materials.[36] Consider the activities of the ATCC. In the early 1980s the agency established its molecular biology program to collect rDNA and vector materials and established one of the earliest bioinformatics programs. With the development of express sequence tags as a primary sequencing technology,

33. Drawn from www.nbci.nlm.gov/Genbank/genbackstats.html (August 2004).

34. Nicholas Wade, "Human Genome Sequence Has Errors, Scientists Say," *New York Times,* June 11, 2002, p. 4.

35. Kristen Philipkoski, "Error in Genome Done on the Fly," *Wired,* April 20, 2000 (www.wired.com/news/technology/0,1282,35814,00.html).

36. Blaine (2003); Hay (2003).

the agency formed an alliance with Craig Venter's Institute for Genomic Research to distribute cDNA clones to researchers. Indeed, in moving to a state-of-the-art facility under the leadership of Ray Cypess in 1998, the ATCC developed a shared facility for research in bioinformatics and the distribution of electronic genomic information. Similarly, the Jackson Laboratory of Bar Harbor, Maine, has built on its traditional role as the world's largest research institution and distributor of research mice to establish its leadership in the characterization and bioinformatics of the mouse genome.

Despite these accomplishments, the revolution in information technology continues to create strong challenges for BRCs and policymakers in the United States and abroad. Effective integration of electronic and biological resources requires mechanisms to ensure the authenticity of materials and the development of policies governing access to data and information. Although the centers have a historical advantage in collecting, preserving, and distributing materials, they are only beginning to master the knowledge management skills required for expertise in bioinformatics. Moreover, aggressive investment by private bioinformatics firms and the development of databases by individual laboratories suggest that establishing common knowledge resources integrating physical and electronic materials at the lowest possible cost will be difficult. Further, the rapid pace of discovery and technological change in both biology and IT make it difficult for BRCs to decide how to prioritize investments in long-lived knowledge resources. Addressing these challenges requires maintaining a balance between exploiting the new opportunities created by the IT revolution and ensuring the continued supply and integrity of traditional biomaterials.

The quest to achieve such a balance is not unique to BRCs. In the past two decades the role of traditional research libraries has been transformed by information technology. Constrained by limited funds, libraries have had to decide between making capital expenditures for digital information systems or expanding the scope of traditional print collections (box 2-2).[37] In determining how to balance their investments, libraries needed to define their function in the information economy. As living libraries for the life sciences, BRCs confront similar issues as they determine how to best exploit the IT revolution.

Biotechnology: Private Commerce or Public Science?

While the information technology revolution is transforming scientific research, the biotechnology revolution is changing the institutions undertak-

---

37. Nicholson Baker, "Discards," *New Yorker*, October 14, 1996, pp. 64–86.

ing life sciences research. BRCs evolved in an environment in which nearly all research was conducted and funded by nonprofit institutions (primarily universities and government laboratories), but most life sciences research in the United States is now funded by for-profit firms, though the bulk is still *conducted* by nonprofit institutions.[38] Starting in the early 1970s, the potential for commercial exploitation spawned hundreds of research-oriented new biotechnology companies, most located in small geographic clusters in or near San Diego, Silicon Valley, Washington, D.C., and so forth.[39] Private sector R&D investment by both large and small firms has been particularly strong in molecular biology and genomics, which provide tools and targets for research to discover new drugs. Specific investment areas in biotechnology now range from gene identification and gene function to the production of DNA chips for research purposes to proteomics and genomic drug development.

Most private sector investment in the life sciences strikes a balance between basic research for publication in traditional scientific journals and more applied research aimed at obtaining formal intellectual property rights and eventual commercial exploitation. While private firms do not directly benefit monetarily from conducting scientific research, their investments enable them to attract higher-quality researchers, and they become aware of new public sector discoveries more efficiently. Although scientific results are public in the sense that they are published in scientific journals, being aware of results before their formal publication and being able to evaluate the importance of novel scientific discoveries depend on employing researchers who also participate in the scientific community. In other words, private firms that would like to exploit the biotechnology revolution must purchase a ticket of admission that pays for itself in greater R&D productivity and a higher rate of technological innovation.[40]

Consider the wave of partnerships and alliances between biotechnology-bioinformatics firms and traditional pharmaceutical companies. Companies such as Pfizer, Roche, and Merck have established wide-ranging alliances for access to "discovery platforms" with bioinformatics-driven companies such as Affymetrix, deCode Genetics, and Celera Genomics.[41] At the core of each of these alliances is an agreement that the established companies will gain preferential access to data and knowledge produced through the research of the smaller company, even while portions of that research will appear in the pub-

38. National Science Board (2002).
39. Audretsch and Stephan (1996).
40. Rosenberg (1990); Cockburn and Henderson (2001); Stern (2004).
41. Ernst and Young (2000).

Box 2-2. Libraries in the Information Age

Some of the significant challenges facing biological resource centers are reflected in the more general contemporary debate about the fate and future of research libraries. These libraries face a number of problems that are only beginning to be addressed: the rapid expansion of higher education, scholarly research, and library collections, sustained budget pressures, and the emergence of digital information systems.

After World War II, research libraries faced a period of unprecedented growth in their collections. According to the *Science Citation Index,* growth in the number of scientific journals has been dramatic since 1950 and was particularly strong during the 1970s (Holley, 1987). Equally important, collections greatly expanded in scope. The growth in the number of specialized, peer-reviewed journals and an expanded range of books in science and engineering put pressure on most universities to build up their science collections, which had barely existed before the war. Even within the liberal arts, research became more international, and professors at most research universities placed priority on having their libraries expand beyond their traditional collection activities. These changes began to place a severe financial strain on libraries by the mid-1980s as the cost of continued accumulation increased more rapidly than university revenues or library budgets.

More important, the late 1980s ushered in digital information systems. For most libraries the first digital initiative was to move the card catalogue online. While it was seemingly a straightforward process, the idiosyncratic pre–digital era habits of librarians made it difficult, subjective, and riddled with error (Baker, 1996). With libraries already facing financial constraints, the higher-than-expected implementation costs of digital cataloguing reduced the funds available to expand library collections at most universities (California State University, 2003).

These challenges have intensified since the mid-1990s with the arrival of the Internet and the explosion of digital media. Increasingly, libraries are confronting difficult trade-offs between digital and physical storage of material. Who should be responsible for the acquisition of digital content? How should access be regulated? What strategies should traditional publishers pursue in terms of online distribution? What standards should be used? Should existing print collections be digitized in an environment of strict budget constraints?

While research libraries sorted out these challenges, researchers and students increasingly demanded digital content. In addition, digital systems offered potential economies of scale, the possibility of costless replication, and access to a much higher level of information, data, and knowledge at lower cost. Pioneers such as the California State University library system attempted to transfer their collections online. Many publishers, however, were unwilling to meet the library system's stipulations, such as allowing continued access to purchased articles on canceled subscriptions, and they declined to participate. Though substantial expenditures and attention were devoted to transition, the California State system ended up with a hybrid, with only 500 online journals. However, even with such setbacks, the goal of implementing a digital system remains. The challenge is for libraries to identify what services to provide and to invest appropriately. As digital standards and encryption techniques improve, libraries will face fewer objections from publishers when providing online access to journals.

Publishers will still serve a necessary function in selecting content, peer evaluation, authentication, and editing. Libraries will move toward being information aggregators and knowledge managers. Knowledge management in a library setting means working within a national library network to properly catalogue information and data available with minimal redundancy. It involves providing proper online search tools, performing selective collection development, and possibly weeding through collections. Research libraries need to embrace their evolved role in the dissemination of information to scholars, to move from being acquirers of physical resources to being managers of digital information.

While the challenges of research libraries parallel those of BRCs in many respects, BRCs encounter additional complexities. First, unlike traditional library collections, which are primarily paper or computer based, BRC depositories consist of sophisticated storage tanks and laboratory equipment preserving active biological materials. Second, BRCs must maintain an active research program as well as a host of laboratories in order to authenticate biomaterials, a function that has no equivalent in a traditional library system. Finally, while interested parties generally visit a research library to access its collection (or increasingly, access it online), BRCs generally must distribute material to researchers who request it. In short, although the issues confronting research libraries and BRCs, insofar as both are knowledge aggregators, are generically similar, the implications for the core activities of each may be markedly more distinct and nuanced. The impact of digitization on BRCs remains to be seen.

lic domain in scientific publications. Indeed, spurred by concerns that such arrangements might limit access to the results of the Human Genome Project, the Universal Declaration on the Human Genome was issued in 1997 to ensure low-cost and independent access to the raw data resulting from the project.[42]

These interactions among public and private researchers raise several issues concerning control of and access to biological materials. Private firms may have incentives to allow their researchers to publish in the scientific literature, but these companies also are trying to take commercial advantage of the materials and knowledge their researchers produce. For example, companies often simultaneously file patent claims and publish scientific papers on new materials.[43] Even when patents are denied, these companies may limit the availability of the materials involved, excluding potential competitors. With the increased commercial viability of university-based discoveries in the life sciences, schools have begun to exert increased control over biomaterials. Although the contribution of reference materials to BRCs was once standard practice, many universities now mandate that researchers request permission before donating potentially valuable cultures.

These trends raise significant challenges for BRCs. On one hand, companies doing research must be allowed to retain appropriate intellectual property rights to their research; on the other hand, the creation of unconnected private pockets of information slows the overall advancement of scientific knowledge as scientists duplicate efforts and face a potential "patent thicket" when pursuing their own ideas. BRCs have attempted to balance these objectives through various initiatives over the past two decades. First, and perhaps most important, the establishment of the centers as international patent depositories had the (perhaps unintended) consequence of ensuring that all patented biological materials are stored in public sector facilities. Many patent depositories, including the ATCC in the United States, have used their status to encourage the development of materials transfer agreements, easing the licensing and transfer of patented materials under their control. Meanwhile, larger BRCs have increasingly served the long-term biological storage needs of many biotechnology, pharmaceutical, and agricultural chemical companies. Some private repositories are maintained, but many collections are eventually included in the centers' catalogues and so become open to public access.

42. UNESCO,Universal Declaration on the Human Genome and Human Rights, 1997 (www.unesco.org/human_rights/hrbc.htm [May 2004]).
43. Murray (2002).

Although important issues have yet to be resolved concerning the appropriate scope and focus of BRCs in the face of substantial private investments in biotechnology, several trends are apparent. First, leading international centers have come to serve as fair brokers between public and private sector participants in life sciences research. This role has eased some of the tensions that have arisen with the expansion of the private biotechnology sector. Second, the centers are increasingly more than simple repositories; they serve as sophisticated knowledge management organizations. How far this new role will extend is open to debate and will be influenced by policy toward both BRCs and biological materials.

Biosecurity: The Increased Threat from Bioterrorism

The terrorist attacks of September 11, 2001, and the anthrax mailings of October 2001 brought bioterrorism and biosafety issues to the forefront of public policy. However, the threat from bioweapons and their implications for control of and access to biomaterials have been increasingly apparent since the late 1980s. Because of their central function in storing and distributing biological materials, BRCs have become a primary consideration in the design of effective policies for regulating access to biomaterials and their distribution.

Although the Biological Warfare Convention of 1972 created an international framework precluding bioweapons development, the lack of effective enforcement (and the politics of the cold war) meant that development of the weapons and the distribution of potentially lethal biomaterials went largely unchecked through the 1980s. In the past decade, mostly in response to specific incidents and loopholes, more stringent controls have been placed on access to and distribution of biomaterials. Three pre-9/11 legislative initiatives stand out.[44] First, the 1989 Biological Weapons Act extended the scope of biomaterials security regulation to individuals and nonstate organizations, making it illegal to possess, trade, sell, or manufacture a biological substance for use as a weapon. Second, responding to instances of international distribution of anthrax and other dangerous materials, the 1991 Export Control Act banned the distribution of materials that could be used in the manufacture of biological weapons to nations suspected of developing such weapons. Similarly, in response to incidents in which nonscientists were able to acquire dangerous biological agents (such as bubonic plague bacteria) through the ATCC and other sources, the Anti-Terrorism Act of 1996 established the

44. For a useful and complete history of the politics of biomaterials security see Knobler and others (2002).

Select Agents Transfer Program (SATP). Administered by the Centers for Disease Control and Prevention (CDC), the SATP regulates the distribution of thirty-six potentially hazardous biomaterials. Organizations sending or receiving these materials must be registered with the CDC and meet minimum safety standards.[45] But because simple possession of materials such as anthrax organisms has not historically required registration, the implementation of SATP has been hampered by the lack of effective enforcement.[46]

While these regulations have limited unauthorized access to biomaterials, distribution of anthrax and other restricted materials for legitimate purposes has also increased in recent years, the result of public sector research efforts in the fight against bioterrorism. Less than two months after September 11 and the October anthrax mailings, for example, researchers at the Mayo Clinic introduced a new diagnostic test for anthrax exposure, dramatically reducing the amount of time required to identify the toxin.[47] Even before September 11, researchers at institutions ranging from Harvard University to the University of Wisconsin–Madison to the University of California–San Diego were working to identify the biology of anthrax and experimenting with potential cures.[48] As Stephen Ostroff of the CDC noted, "there is a need to expand research involving select agents, not to constrain it. We must bring the best and brightest minds to bear on the development of better vaccines, antiviral agents, antibiotics, and other therapies for exposure to, or illness from, biological agents. To do so, we need to ensure that restrictions on possession or handling of biological agents do not have a chilling effect on the willingness of scientists and research establishments to take part."[49]

Biological resource centers have been at the forefront in attempting to avoid this chilling effect. The costs and risks of distributing biological agents have made most laboratory-to-laboratory transfers infeasible: as early as 1998 Ronald Atlas of the American Society of Microbiology testified, "What has happened is that a number of individuals are simply not shipping. They are not exchanging."[50] As a result, the ATCC and other large centers have

45. Centers for Disease Control, "42 CFR Part 1003: Possession, Use, and Transfer of Select Agents and Toxins; Interim Final Rule," December 2002 (www.cdc.gov/od/sap/docs/42cfr73.pdf [May 2004]).

46. See Ostroff (1999).

47. Kenneth Chang, "New Tools Emerging to Speed Anthrax Detection," *New York Times,* November 8, 2001, p. B9.

48. Sam Perdue and Laurie Doepel, "Researchers Discover Secrets of Anthrax's Killer Toxin," October 2001 (www.niaid.nih.gov/newsroom/releases/anthrax.htm [May 2004]).

49. Ostroff (1999, p. 24).

50. Atlas (1999, p. 60).

become de facto authorized distributors of these materials. Although there have been occasional criticisms of BRCs when materials have been unintentionally distributed to unauthorized personnel, these repositories are likely to be more effective than lab-to-lab exchanges in managing the registration process, tracking shipments, and meeting other security requirements. By taking on the responsibility of handling restricted materials, the centers allow researchers to benefit from economies of scale in satisfying new regulations without unduly restricting legitimate exchange of materials or research.[51]

These concerns have become much more pressing since September 11. Despite widespread agreement that a new and more comprehensive policy response is warranted, there remains little consensus about how best to balance biosecurity needs with the desire to maintain open exchange of materials, particularly for labs whose research directly contributes to antiterrorism efforts. Effective policy here must bridge the historical divide between national security agencies and the scientific research community. Because of the historical role of BRCs in mediating between both groups as well as their continuing status as the most effective handlers of sensitive biomaterials, the centers are likely to be crucial in developing and implementing biomaterials policy in the future.

Biodiversity: Preserving Nature

The final emerging challenge for BRCs results from their unique function in addressing biodiversity issues. In the past decade there have been increasing calls for both the systematic preservation of species, including microorganisms, and the establishment of sovereignty rights over naturally occurring biological organisms. But although biodiversity is an agreed-upon goal, experts disagree on the feasibility of preserving organisms, particularly in the current policy environment. Many argue that because the cost of preserving all organisms is prohibitive, priority for deposit in a BRC should be reserved for those materials with demonstrated scientific value. As noted by Arnold Demain of MIT, biodiversity is "an interesting cause, but where would the money come from? The goal is sort of crazy because economically it would be impossible. . . . It would be nice if all cultures could be made available, but it's never going to be possible."[52]

However, many researchers would argue that the scientific value of mate-

51. Simione (2003).

52. Arnold Demain, interview with author, MIT Department of Biology, December 5, 2000.

rial cannot be determined in advance and that broad-based conservation efforts are likely to yield important scientific findings themselves. For example, the development of PCR by Kary Mullis in the 1980s depended crucially on the availability (at the ATCC) of *Taq* polymerase, a microbe derived from the extremophile *Thermus aquaticus*. When *Thermus aquaticus* was originally deposited in the ATCC in 1969, it had no particular scientific merit and none was anticipated. So BRCs play a special role in preserving biodiversity and will be significant participants in any systematic efforts to catalogue naturally occurring materials. Specifically, compared with individual laboratories or for-profit culture distribution centers, BRCs have historically preserved the most diverse range of materials.

Biodiversity concerns raise new and complex policy challenges for BRCs in both the United States and Europe and for emerging economies.[53] In 1992 the Convention on Biological Diversity (CBD), otherwise known as the Rio Summit, attempted to make significant strides toward persuading the scientific community to integrate the goal of biodiversity into its practices. Although past agreements had focused on one aspect of diversity at a time, the 1992 proceedings pursued a holistic approach. The summit's aim was to create a comprehensive framework that would help nations begin initiatives and enforce programs to preserve natural biodiversity. The primary policy result of the CBD was agreement that individual countries have the right to set rules and regulations governing distribution of naturally occurring biological materials. Essentially, by linking sovereign nations to responsibility for biological conservation, the CBD framework allocated property rights over biological organisms to create incentives for their effective care and preservation.

From the perspective of BRCs, however, the agreement has created a new and uncertain set of responsibilities for deposit and storage of biological materials. When researchers deposit material in a center, the CBD requires that they certify that they have been granted access to and control of the material by the country it came from. Such requirements have created quandaries for BRCs and raised their own questions. How should each country formulate its rules and regulations so as to exert its rights under the agreement? Are national culture collections a necessary component of compliance with the CBD? What resources should be allocated to establish compliance in developing and emerging economies? Each of these questions must be addressed for BRCs to respond and create policies to preserve global biodiversity.

53. Staley (2003).

Conclusion

Biological resource centers have been productive and important in the development and growth of life sciences research during the past half century. From their early impact on virology to their crucial part in addressing cross-culture contamination in the 1970s to their current leading role in facilitating the exchange of materials protected by formal intellectual property agreements, the centers have established themselves as vital to innovation in life sciences research. However, the discussion so far has neither identified why BRCs are able to perform these tasks to improve the environment for life sciences research nor whether establishing effective centers requires intervention by government or the research community. Such a discussion requires a systematic analysis to identify the sources of the benefits arising from BRCs and their potential costs.

3

Climbing atop the Shoulders of Giants: BRCs as Cumulative Knowledge Hubs

B y helping establish norms and procedures for authenticating and exchanging biological materials, biological resource centers have been crucial in the development of the life sciences in the past half century. Though the centers are invisible to the public, scientists are keenly aware of their impact, from their early influence on virology to their crucial part in addressing cross-culture contamination in the 1970s to their current leading role in creating an international BRC network. As highlighted by a comment that echoed through the author's interviews with depositors and users of BRC materials: "They are like the Library of Congress for biological materials."

This chapter evaluates the contribution of institutions like BRCs from an economic perspective. The analysis examines the critical role of cumulative knowledge in furthering scientific, technological, and economic progress. By allowing today's researchers to build on ideas and knowledge developed in the past, cumulative knowledge allows researchers to stand "on the shoulders of giants," in Isaac Newton's words, generating spillovers between research generations.[1]

---

1. Though certainly not the first example, Newton's recognition of cumulativeness is (famously) recognized in his classic 1676 letter to scientific rival Robert Hooke in the context of a dispute over the nature of light: "What DesCartes did

Progress can be enhanced through establishing "cumulative knowledge hubs."[2] A cumulative knowledge hub is an institution or system enabling knowledge transfer among researchers through centralized deposit and access facilities. Knowledge hubs perform three central functions: acquisition of materials (including physical materials, databases, and other media), certification of the fidelity of the materials, and provision of long-term access to certified materials to researchers.

Compared with alternative knowledge distribution systems, knowledge hubs promote cumulative progress. This chapter illustrates how BRCs serve as more than simple materials repositories. As knowledge hubs for life science research, they contribute to lower research costs and increased research productivity. These contributions are at the heart of the unique role they have played in the growth of life sciences research over the past several decades. By offering a systematic way to assess BRCs' contributions, the knowledge hubs framework provides a foundation for policy assessment, allowing consistent evaluation of the various policy issues facing these institutions. Consider the BRC funding environment. While some BRCs rely almost exclusively on government financing, others derive substantial income as patent depositories and from user fees. What part should government financing play in the future of BRCs? What changes to current funding policy should be implemented for the centers and the life sciences community to thrive?

The knowledge hubs framework also illuminates issues relating to access to and control of research materials and resources. Should publicly funded researchers publishing studies relating to new biological materials be required to provide access to those materials? To what extent should depositing materials in a BRC serve to comply with access regulations? Should these rules depend on whether the materials have potential security and safety implications? Should researchers be given incentives to use the centers' biomaterials? If so, why?

was a good step. You have added much several ways, & especially in taking ye colours of thin plates unto philosophical consideration. If I have seen further it is by standing on ye sholders of Giants."

2. The concept of knowledge hubs has become popularized in the last several years in a number of separate contexts, ranging from the use of knowledge management systems within individual companies (Mach and Owoc, 2001), to mechanisms to facilitate electronic commerce (Kaplan and Sawhney, 2000) to means to enhance location-specific knowledge accumulation(Garrett-Jones, 2004; Owen-Smith and Powell, 2004). Recent research has emphasized the crucial role played by network hubs in the structure and performance of networks (Barabasi, 2002), with particular focus on the research networks (Powell, Koput, and Smith-Doerr, 1996; Powell and others, 2004). The specific framework developed here links the performance and structure of scientific and commercial research networks to specific institutions that serve as centralized knowledge brokers.

The knowledge hubs framework also provides a guide to emerging BRC policy issues. How should federal policy toward BRCs and the control of physical biological materials adapt to the rapid digitization of life sciences research? What are the most effective policies for addressing the development of the growing international network of BRCs, many of which would like to rely on U.S. expertise and leadership in this area?

The remainder of this chapter discusses the role of cumulative knowledge in scientific progress and economic growth and the challenges in ensuring the accumulation of knowledge over time. It then defines and identifies the contributions of knowledge hubs to that process and evaluates BRCs in light of the knowledge hubs framework.

Cumulative Knowledge: The Driver of Economic Growth and Scientific Progress

The economic impact of BRCs derives from the importance of scientific research and innovation as well as the effects of institutions and policy on long-run growth and productivity. The importance of institutions supporting scientific research and innovation in encouraging long-term growth has been recognized and incorporated into federal policy for more than half a century. In 1945 Vannevar Bush's *Science: The Endless Frontier* presented a compelling assessment of the importance of basic research funding and support for scientific progress in economywide prosperity and security.[3] Since then, as the United States has developed and exploited the most advanced scientific infrastructure in the world, economists and policymakers have come to appreciate the centrality of science, technology, and innovation in ensuring national security, maintaining competitiveness, and sustaining long-term economic growth.

The importance of scientific and technological progress in raising living standards over the past 200 years has been appreciated at least since the research of Robert Solow.[4] However, the mechanisms by which science and technology operate and matter for prosperity are less clearly appreciated. In the past two decades, economists have come to appreciate that economic progress depends in large part on the environment for innovation and the effectiveness of institutions supporting cumulative advancement.[5] Specifically, sustained

3. Bush (1945).
4. Solow (1957).
5. Nelson (1993). More generally, Rosenberg (1963) was the first to identify the critical interdependencies contributing to the innovative capacity of an economy, highlighting the role of knowledge spillovers among closely related economic sectors and the presence of institutions supporting technical advance.

growth in productivity depends on the vitality of the sectors and industries with a strong connection to science and particular scientific disciplines.[6]

These observations are at the core of knowledge-based economic growth. As described by Paul Romer and other "new growth" economists, economic growth depends on the ability of knowledge (or ideas) to produce valuable innovations serving current needs by building on the findings and discoveries made by scientists and researchers in the past. To serve as a foundation for long-term growth, scientific research and technological progress must continually stand on the shoulders of earlier researchers.[7] If the knowledge pool stagnates, diminishing returns set in and growth halts. However, if knowledge is cumulative and discoveries are continuous, diminishing returns may be held at bay, allowing sustainable long-term growth.[8]

The Challenge of Cumulative Knowledge Systems

Economists and policy analysts generally assume that the production of scientific and commercial knowledge results in cumulative knowledge. However, the mere production of knowledge does not guarantee that others will be able to exploit it, since the conditions that lead to cumulative progress are by no means guaranteed. Researchers must both be aware of the existence and believe in the truthfulness of prior discoveries; if not, they are forced to reinvent the wheel in their own research. Uncertainty necessitates costly reverification and reinterpretation, reducing research productivity. Without a system to ensure cumulativeness, the ability of follow-on researchers to benefit from earlier discoveries may be sharply limited.[9]

The scientific method facilitates the achievement of some of the prerequisites for the effective accumulation of knowledge. The crux of the scientific method is that all discoveries must undergo independent verification and replication and that knowledge should be shared by members of the scientific community. If these practices operate effectively, researchers within a field can confidently access knowledge produced by others, facilitating cumulativeness.[10] For example, James Watson and Francis Crick's discovery of DNA's double helix structure in 1953 both depended on previous experiments (such as the X-ray crystallography work of Rosalind Franklin) and laid the foundations for the revolution in molecular biology studies.

6. Rosenberg (1974); Adams (1990); Marburger (2002).
7. Romer (1990); Grossman and Helpman (1991); Jones (1998).
8. Romer (1990); Jones (1995).
9. Mokyr (2002, pp. 7–10).
10. Dasgupta and David (1994).

However, the integrity of the scientific process is not a given. Scientists have incentives to focus on their publication record and potential commercial returns on their discoveries. While publication requires some disclosure, most scientific research cannot be replicated simply by reading the literature. Without access to key materials, equipment, and methods, all members of a scientific community cannot reproduce published scientific result, and so the results have limited cumulative value. In addition, individual scientists may allow only researchers with whom they interact in a tightly knit research network to have access to important materials or methods.[11] Indeed, in their race with Linus Pauling to discover the structure of DNA, Watson and Crick took advantage of their preferential access to the crystallography images in Maurice Wilkins's lab; Pauling's inability to visit England during the early 1950s may have been decisive in determining who got credit for the discovery.[12] Finally, while professional stature, prizes, and (occasionally) wealth provide strong incentives for researchers to compete for discoveries, the scientific system offers researchers few incentives to authenticate others' findings.[13]

Ironically, though the effectiveness of cumulative knowledge depends on mechanisms for ensuring the integrity of scientific and technological knowledge, few incentives exist for individual researchers to develop or maintain such mechanisms. Instead, step-by-step innovation is inextricably linked to the presence and effectiveness of real-world institutions and policies.[14]

The Nature of Cumulative Knowledge Hubs

Although many types of institutions help ensure the integrity of cumulative research, institutions or systems offering centralized deposit and access facilities for research materials and resources—knowledge hubs—are particularly important. Knowledge hubs perform three main functions: acquisition, authentication, and the provision of access. From straightforward examples such as open access libraries and scientific journals to more subtle examples such as standard-setting bodies, knowledge hubs allow scientific and technological communities to develop and extend the range of knowledge that members draw on in their own research. To understand the significance of knowledge hubs, it is useful to compare them with alternative mechanisms for accumulating knowledge.

First, in the absence of a way to access the materials needed to draw upon

11. Powell, Koput, and Smith-Doerr (1996); Rosenkopf and Tushman (1998).
12. Watson (1968).
13. Merton (1973).
14. Nelson (1993); Mokyr (2002).

existing knowledge, individuals (or individual organizations) are likely to develop private materials collections and an individualized knowledge base. Many high-technology companies, from Microsoft to Pfizer, maintain proprietary internal databases to capture and leverage the knowledge produced by their researchers and employees.[15] Even for basic research within universities, control over key materials and the details of earlier results often are not fully disclosed; as a consequence, researchers tend to build on their own research, unwilling to incur the costs associated with replicating the work of others in their field.[16]

Even in the absence of specialized institutions it is possible for researchers to exchange materials (including physical materials, databases, and equipment) with each other through the peer-to-peer network. In these transactions, a researcher who wants to draw on a discovery must negotiate with the initial discoverer; for example, the researcher may have to offer the discoverer coauthorship of a paper in exchange for access to lab data or cell lines. Of course, in a peer-to-peer network it is impossible to require researchers to exchange materials, and original discoverers may be reluctant to offer access to another researcher who might pursue experiments that undermine the value of their initial work.

It is also possible for informal brokers to emerge in a peer-to-peer network, facilitating transactions in the market for ideas.[17] While such brokers perform some of the same functions as knowledge hubs, they are limited by the size of their personal network and their incentives. In addition, the connections enabled by a broker may be terminated if, for example, the broker retires or his or her interests change.

While this book's assessment of biological resource centers focuses on institutions that are predominantly in the public sector (either directly controlled by government or organized as nonprofit associations governed by scientific societies), knowledge hubs may either be for-profit, not-for-profit, or government organizations. Although for-profit hubs are likely to manifest some of the benefits of effective knowledge hubs, they are less likely to offer sufficient incentives for engaging in difficult-to-monitor certification activities, long-term preservation of materials, or marginal cost pricing to enable

15. For example, Microsoft requires all software engineers to contribute solutions to problems encountered on each project to a database that can be accessed for all future projects (Cusumano and Selby, 1997).

16. It is important to note that, by the very nature of the scientific method, private collections cannot be authenticated because it is impossible to verify the integrity of one's own knowledge and discovery.

17. Lamoreaux and Sokoloff (2001); Gans and Stern (2003a).

Figure 3-1. *The Knowledge Hub Network Structure*

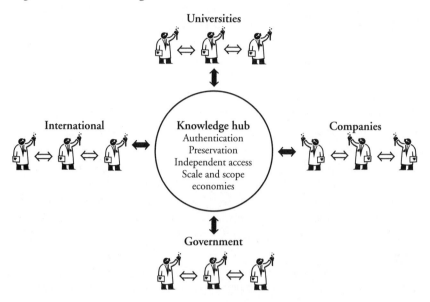

the widest possible access. More generally, since the benefits arising from a knowledge hub derive from its ability to certify knowledge without reducing the credit received by those associated with an initial discovery, for-profit organizations may find it difficult to maintain the level of "invisibility" required for effective functioning. In other words, for-profit organizations, by their very nature, will be predisposed to claim discoveries as their own and to attempt to profit from the economic benefits that such discoveries create. On the other extreme, while many knowledge hubs are sponsored and operated directly by governments, government control may hinder the effective operation of hubs whose users span the international scientific and technological community.

Relative to alternatives such as peer-to-peer networks, knowledge hubs offer at least four advantages in fostering cumulative knowledge (see figure 3-1). First, by confirming the fidelity of materials underlying a discovery, a hub certifies the validity of the findings. In the absence of certification, mistakes can (and almost certainly do) filter into the knowledge stock. The costs of such mistakes can be substantial because all knowledge that builds on false findings must be reverified and the process of reconstructing knowledge often can be nearly as expensive as the initial process of discovery or invention. Moreover, when certification is not routinely available, even the potential for mistakes forces researchers to reverify results before proceeding. Second, knowledge hubs can store materials indefinitely. Knowledge that may have little value when first discovered may be exploited effectively when complementary technology and developments reveal the full value of the initial piece of knowledge. Third, while peer-to-peer transactions often suffer from incentive and contracting problems, a hub allows for independent access by later researchers. Independent (and low-cost) access allows certified materials and knowledge to become available to a wider audience. Fourth, hubs achieve economies of scale and scope in aggregating, authenticating, and distributing knowledge, in part because the efficiency of the process may improve substantially over time. Overall, the existence of an effective knowledge hub greatly increases a technological or scientific community's capacity to accumulate knowledge. At its core, a knowledge hub amplifies the impact of a piece of knowledge by stocking the materials on which the knowledge is based in a collection upon which future researchers draw.

Knowledge Hubs for the Life Sciences

My evidence, based on eighteen months of field study, suggests that, as economic institutions, BRCs are key knowledge hubs in the life sciences. Indeed, this role was repeatedly emphasized in interviews. As one respondent said, the centers are "the infrastructure that glues together many fields of investigation and investigators themselves by providing . . . quality materials." To assess this claim more clearly, consider how BRCs manifest each of the advantages of a knowledge hub.

Authentication

First, through their certification function, biological resource centers have emerged as honest brokers in the life sciences. As scientific progress becomes more mingled with technological innovation and its associated economic rewards, the integrity of the life sciences has come to rely more and more on unbiased institutions to certify research results published in scientific studies or in the patent record. While the centers do not fully replicate experiments published in scientific journals, all materials incorporated into BRC collections undergo review and testing to establish the identity and biological viability of the material. As the mission statement of the American Type Culture Collection states, "We provide the authenticated biological materials . . . you decide how to use them in your research."[18] BRCs therefore provide the

18. American Type Culture Collection (ATCC), www.atcc.org/About/About/ATCC.cfm (July 2002). Some BRCs, including the ATCC, offer a classification system allowing researchers to evaluate the degree of confidence associated with specific deposits.

Box 3-1. The HIV Virus Debate

When material exchanges occur through the peer-to-peer network, credit for discoveries and the validity of findings can become clouded by uncertainty, tarnishing the reputations of venerable scientists and slowing the progress of an entire research community.

The race to discover the cause of AIDS involved an intense battle between French researchers at the Institut Pasteur and Robert Gallo's lab at the National Cancer Institute. Though the French team first isolated the correct virus, laboratory-to-laboratory material exchanges resulted in nearly a decade of confusion about the precise nature of the virus and the allocation of credit for its initial discovery, damaging Gallo's reputation and delaying critical AIDS discoveries.

Gallo first identified human retroviruses in the late 1970s. With the appearance of AIDS, interest in these microorganisms increased dramatically. During the competition between the Gallo and Pasteur labs, the labs exchanged samples several times. Most notably, in 1984 the French team sent Gallo samples the team had used to identify the LAV virus as the key culprit causing AIDS so that Gallo's lab might compare LAV to the virus the lab had identified, HTLV-3B. Almost immediately, a dispute erupted. While the French team claimed that LAV was distinct from HTLV-3B, Gallo claimed that he and the French team had identified the same organism. Unable to resolve their differences on the distinctions between the organisms and issue of priority, the National Institutes of Health and French officials agreed to accept joint responsibility for the discovery, simultaneously publishing papers and sharing ownership of a patent for the AIDS blood test (Shilts, 1993).

---

means for efficient scientific replication and allow for improved allocation of credit for results (box 3-1).

Though seemingly straightforward, authentication is critical to effective life sciences research. Consider the early history of peer-to-peer networks and the misidentification and contamination scandals arising from HeLa and other early tissue cultures. The consequences of such events are far-reaching. Not only does misidentification cast a cloud over the findings of current researchers (and over the careers of those whose results are under suspicion), but the resulting confusion and uncertainty also place a longer-term cost on progress. Researchers must painstakingly reestablish the validity of findings in order to design and implement new research. Even in the modern era,

During the next decade, controversy continued among researchers about the typology of the AIDS virus and how to allocate credit (and associated patent royalties) between Gallo and the Pasteur Institute for the discovery. Eventually, it was discovered that the samples of both labs had been contaminated at some point, subjecting the discoveries of both to question and leading to confusion and uncertainty about the validity of follow-on research (Crewdson, 1995; Subcommittee on Oversight and Investigations, 1995). Further studies concluded that Gallo's initial research results were based exclusively on samples from Pasteur (thus stripping Gallo of his status as codiscoverer and raising questions of scientific integrity), although it remains unclear whether this error was intentional or inadvertent due to the contamination of samples. As noted in the final report of the House Subcommittee on Oversight and Investigations, "the consequences for HIV research were severely damaging, leading, in part, to a corpus of scientific papers polluted with systematic exaggerations and outright falsehoods of unprecedented proportions" (Subcommittee on Oversight and Investigations, 1995).

A good portion of the blame for this debacle can be placed with the routine practice among scientific researchers of informally exchanging materials. Lacking any consistent or standard mechanism for authenticating and certifying cultures, such exchanges can lead to confusion about the precise origin and properties of materials, or worse, to contamination. Beyond the personal impact this might have on researchers whose results may be rendered invalid, such uncertainty may undermine the integrity of scientific research generally. The degree of confidence in follow-on research is limited by the lack of controls over the materials tested, reducing the progress possible through the scientific method and wasting scarce resources on later research that is invalid from the start.

BRCs seem to be the most aggressive and systematic developers of technology and tools for eradicating subtle problems, such as genetic drift (changes in the genetic structure of research organisms over time) and mycoplasma infection (a microbial invasion that influences research findings in subtle ways).[19] By offering authenticated materials, resource centers allow researchers to confidently exploit materials used by earlier scientists, avoid needless and costly duplication, and so increase research productivity.

Certification also enables more effective standardization of biological models and experimentation procedures. The value of a biological material or

19. Blaine (2003, p. 24).

Box 3-2. The Anthrax Search

In the absence of effective certification and materials standardization procedures, accurate identification of biological materials can be seriously hampered. Authentication and certification of biomaterials is a basic BRC function. Without a central entity responsible for cataloguing biomaterials,
identifying their origins becomes difficult if not impossible. Consider the
anthrax mailings of October 2001. These bioterrorist attacks killed five people, infected twenty-two others, and caused significant government and commercial disruption. However, despite the most intensive effort in history to
identify the origin and typology of a particular bacterium, more than seven
months was required to issue a detailed genetic analysis that identified a specific U.S. Army research lab as the most likely initial source (Weiss, 2002).

Two elements of the pre–September 11 biological materials control system
contributed to the slow progress in the anthrax investigation. First, laboratories had not been required to register possession of anthrax organisms
although registration of distribution was required as of 1996. Beginning in
November 2003, registration is required by the Public Health Security and Bioterrorism Preparedness and Response Act of 2002. As of September 2004, however,
no laboratory had issued a certificate of compliance, and disputes over the distribution of biomaterials are likely to continue going forward. (See press release of
November 12, 2003, from Sen. Joseph Lieberman at Govt-aff.senate.gov/index.
cfm?FuseAction=PressReleases.Detail&Affiliation=R&PressRelease_id=571&
Month=11&Year=2003, as well as William J. Broad's article, "I Beg to Differ: In a
Lonely Stand, a Scientist Takes on National Security Dogma," *New York Times*,
June 29, 2004, p. F3.)

model generally increases with its use by researchers, since repeated use tends
to reduce the experimental uncertainty associated with a given investigation.
This results in a positive feedback loop, increasing the use of a small number
of biological models in an ever greater number of experiments.[20] Moreover,
when new strains are not introduced into centralized collections, there is no
mechanism for maintaining up-to-date knowledge on all available strains,
specific variants of strains, and the control and ownership of strains (box

20. This dynamic is similar to the process of standardization and lock-in found in many
other high-technology areas, such as computer software and telecommunications equipment.
Economists have paid increasing attention to the impact of "network externalities" over the last
decade, developing implications for antitrust and intellectual property policy (Shapiro and Varian, 1999).

At the beginning of the anthrax investigation, investigators could not even identify the laboratories that may have held anthrax bacteria for months after the initial attacks. Even today, it is difficult to know whether investigators have assembled an exhaustive list of all sources of laboratory anthrax stock. The historical lack of a registration requirement, combined with the current scrutiny of those possessing stock (particularly if lax controls have been in place), may lead some laboratories to be less than forthcoming about their holdings.

Second, as noted above, there has been no historical registration require-ment for new anthrax strains, making it very difficult to trace the origins of the strain used in the October mailings. One of the keys to identifying the origin of the anthrax used in the mailings lay in distinguishing that particular strain from the hundreds known to exist in laboratories throughout the world. After genetic sequencing of the anthrax used in the mailings was com-pared with known samples, scientists were able to determine that the anthrax had its origins in the extensively used Ames strain and that this particular variant was closely related to a strain found at the Fort Detrick laboratory col-lections.

Overall, the absence of a registration requirement for possession of anthrax stock and for development of new anthrax strains greatly limited the ability of investigators to conduct an efficient investigation into the origin and specific typology of the anthrax used in the mailings. Requiring deposit and registra-tion with BRCs of selected new materials strains will help prevent future acts of biological terrorism, greatly improve the efficiency of investigation in future pathogenic outbreaks, and facilitate the public health response to such outbreaks.

3-2). Certified biological reference materials standardize experimentation procedures, increasing the scientific value of results.

BRCs create a common database from which to draw materials. For exam-ple, Nancy Wexler's deposit of more than 1,000 samples at the Coriell Insti-tute, which were crucial for identifying the Huntington's disease gene, "forged a scientific collaboration based on sharing of ideas and data that was unparalleled in biomedical research."[21] By documenting the use of materials by other researchers (through the standardized use of accession numbers and the like) and actively monitoring trends in the use of materials by researchers, BRCs increase the strength and effectiveness of research networks and encourage the appropriate use of standardized biological materials.

21. Beck (2001).

Box 3-3. Brock's Unlikely Bacteria

BRCs preserve material whose value, while unclear and unknown at the time of deposit, may lead to dramatic advances in scientific productivity itself. In 1967 Thomas Brock of the University of Wisconsin discovered thermophilic bacteria at Yellowstone National Park. The story of the development and impact of PCR is widely documented (for a particularly useful account see Rabinow, 1997). This enzyme, *Thermus aquaticus*, occurs naturally in hot springs, remains stable at up to 95 degrees Celsius, and retains its enzymatic properties at much lower temperatures. Although a use for this material was unclear at the time, Brock hoped to make it available to researchers indefinitely. He therefore deposited the bacterium with the ATCC. Indeed, it was nearly twenty years before Brock's deposit would change modern biology forever.

Kary Mullis, at that time a researcher at Cetus Corporation, conceived the idea of polymerase chain reaction (PCR) without specifically knowing how to implement it. His insight was that isolated strands of DNA could be amplified by exploiting the innate replicability of DNA material. Creating PCR consists of three steps. First, DNA is denatured into single strands by heating it to over 90 degrees Celsius. Second, primers are added and the temperature lowered during an annealing phase that binds primers to each strand. Finally, when a DNA polymerase is added, complementary nucleotides match with each other, and the construction of a replicant DNA strand commences (thus exploiting the process that occurs during natural reproduction). And DNA replication through PCR is iterative: even with cycles of over a minute, millions of copies can be generated in less than an hour.

To translate this idea into a revolution, however, required identifying a DNA polymerase that would not lose its enzymatic properties during heat-

Preservation

Biological resource centers collect, characterize, and maintain a richer and more varied collection of biological materials, particularly those whose value is not initially obvious, than alternative organizational forms. Kary Mullis's ability to develop the extremely influential polymerase chain reaction technique in the late 1980s, for example, relied heavily on the ATCC's long-term storage of a strain of extremophile, *Thermus aquaticus*, whose value could not have been predicted at the time or until many years after initial discovery (box 3-3).

In contrast, the dispersed peer-to-peer network results in tremendous replication with little incentive for any one laboratory to maintain the full

ing. In the early attempts to realize the PCR technique, the problem faced by Cetus researchers was that fresh DNA polymerase needed to be added for each cycle, substantially slowing the process and making it subject to error and potential contamination. In their search to refine the technique, they turned to the ATCC. Though Brock's deposit had been mostly forgotten, the ATCC had preserved *Thermus aquaticus*. Amazingly, the DNA polymerase of *Thermus aquaticus*, dubbed the *Taq* polymerase, was ideally suited for the rapid heating and cooling required of PCR. The combination of Mullis's idea and *Taq* resulted in perhaps the single most powerful research tool in modern biology and biotechnology, with applications ranging from genomic sequencing to genetic fingerprinting. Indeed, shortly after the unveiling of *Taq*-enabled PCR in 1985, scientists shifted their research priorities to take advantage of the new method; as a result, *Taq* was declared Molecule of the Year in 1989 by *Science*, the principal journal of the American Association for the Advancement of Science, and Mullis was awarded the Nobel Prize in 1993.

Without the preservation activities of BRCs such as the ATCC, *Thermus aquaticus* and its DNA polymerase may not have been easily available to Cetus researchers. By characterizing and cataloguing the properties of the sample, preserving it for nearly two decades, and allowing independent access to the ATCC collections, the ATCC dramatically reduced the cost of the development of PCR. Indeed, without the sample, Cetus researchers may have introduced the much more inefficient technique they had developed before their acquisition of *Taq*. Simply by following its own policies, the ATCC facilitated the transformation of a seemingly unknown thermophile into one of the most important discoveries in modern biological research.

range of materials of potential use to researchers at other laboratories. More important, the maintenance of materials in the peer-to-peer network depends on individuals, raising the possibility that the materials will be lost because of the retirement or inattention of culture curators. For example, in early 2002 three private university collections were identified as "orphans" available for a new storage site; two of them were classified as defunct by July 2002.[22]

At the same time, the intellectual property held by for-profit laboratories remains in force for only a modest time (often less than the time between its

22. United States Federation for Culture Collections, "What Are Endangered Culture Collections?" (http://methanogens.pdx.edu/usfcc [May 2004]).

initial characterization and greatest potential use), leaving for-profit labs with few incentives to indefinitely maintain the widest range of materials. Indeed, for-profit distributors of biological materials have often succumbed to cherry-picking, focusing on a narrow range of materials offering high margins and low storage costs. Because these companies are less likely to capture the full value of long-term variety, nonprofit BRCs are critical to collecting, characterizing, and distributing biological materials.

BRCs also preserve a permanent record of the flow of biological materials among researchers. By formalizing and documenting the exchange and use of these materials, the centers are indispensable in managing biological knowledge.[23] For example, the use of BRC materials allows for the rapid assessment of the originality of claims made in research papers and patent applications (see box 3-1). By reducing the costs associated with the assessment of claims, the centers increase the productivity of research activities. In some circumstances, this documentation becomes a critical national security resource. The recent anthrax investigations, for instance, have been severely impeded by the absence of a centralized database of exchanges of biological materials. Compared with peer-to-peer exchanges or for-profit laboratories, BRCs are recognized for their ability to systematically track the flow of biological materials (see box 3-2).

Independent Access

Because BRC materials are equally accessible to all members of the scientific and technological communities, the centers facilitate independent access to the results of scientific research. In non-BRC networks, access to source materials depends on the goodwill of researchers who maintain active cell cultures in their laboratories; however, goodwill is difficult to maintain when researchers are competing with each other to establish new findings or when a particular experiment may cast previous findings in an unfavorable light. Alternatively, for-profit characterization (or materials description and authentication) and distribution companies often find it in their private interest (though not in the social interest) to arrange for exclusive access to their databases and materials. Recent controversies over the "ownership" of the results of the Human Genome Project are but the most visible in the ongoing wars over access to biological materials and data.

Building on earlier research in the peer-to-peer system may involve protracted negotiations with the original researchers, for example, about crediting coauthors or recognizing intellectual property claims. Though access to

23. Blaine (2003, pp. 17–24).

research materials required for replication is at the heart of the scientific method, the incentive of individual scientists to grant access is limited. Even after results are published (and perhaps because they are published), researchers may hold up efforts by others to gain access to materials, both to extend their own research lead and to avoid detailed investigation of their conclusions.

In contrast, BRCs certify the fidelity of the materials that they distribute, severing the direct tie between the researcher associated with an initial discovery and those wanting to build on the discovery. For example, during the HeLa scandals of the 1960s, only the ATCC was willing to undergo the rigorous review that established the seriousness of HeLa contamination. In recent years BRCs have invested heavily in promoting independent access, even for materials protected by intellectual property rights.[24] In their function as international patent depositories, they have established sophisticated protocols for transfer agreements, facilitating the exchange and distribution of patented and nonpatented biological materials (box 3-4). Overall, as leaders in life sciences knowledge management, the centers provide a mechanism for diffusing materials (and the knowledge associated with them) to a wider audience over a longer period of time.

Economies of Scale and Scope

As living libraries that continuously collect material developed by the scientific community, biological resource centers may achieve substantial economies of scale and scope. Many BRCs maintain large, varied, and balanced collections, and they are more likely to make the investments necessary to increase the quality and reduce the cost of accessing biological materials. Over the past decade the ATCC, Germany's DSMZ, the Coriell Institute, the Jackson Laboratory, and other institutions have established positions of global leadership in specific materials and collections. The growth of these large, yet cohesive, collections has coincided with a substantial commitment to high quality for each activity under their domain. They have thus been able to offer access to a larger, more diverse, and more balanced collection at a lower cost than the alternatives. These economies are reflected in the increasing use of nonprofit BRCs for private collections (for example, by private pharmaceutical and biotechnology companies) and in the use of the centers as official international patent depositories. In contrast, in the more dispersed peer-to-peer network, duplication abounds across laboratories and there are few incentives to maintain the highest levels of quality or the broadest portfolio. Another advantage of a broad collection is that it allows acces-

24. Cypess (2003c, pp. 36–42).

Box 3-4. The Budapest Treaty

As international patent depositories, BRCs facilitate independent access to biological materials even when the materials are protected by intellectual property laws. As patenting of biological materials began to increase dramatically during the 1970s, the disclosure requirements of the patent process began to conflict more and more with the requirements of living materials. Providing disclosure on a living organism requires that a specimen of the organism (or its genomic sequence) be made available through the patent record. Indeed, in part because of the lack of a legal framework for ensuring disclosure, the patentability of living organisms in the United States was not clear until the 1980 Supreme Court ruling in *Diamond* v. *Chakrabarty*. Throughout the 1970s, different countries and patent offices began to develop different rules about patentability, with the potential for substantial inconsistency and inefficiency for those seeking global patent protection.

In conjunction with patent office policymakers, biological resource centers around the world came together during the 1970s to develop a systematic and consistent international framework for governing the disclosure and preservation of patented living materials. The Budapest Treaty on the International Recognition of the Deposit of Microorganisms for the Purposes of Patent Procedure was introduced in 1977 and implemented in 1981 in the world's major patent regimes. At the heart of the treaty are regulations governing international patent depositories. Within each country that is party to the treaty, the grant of international patent depository status to a domestic BRC

---

sion of materials whose initial value is uncertain; a wider collection allows the life sciences community to maintain its option to use various biological materials. Particularly for the evolving bioinformatics field, exploiting such economies is crucial to supporting the increasing use of materials by life sciences researchers.

Overall, BRCs serve as the main knowledge hubs for the life sciences. Compared with peer-to-peer exchanges, BRCs make research materials and results more available to other researchers for follow-on research. At a slightly more subtle level the centers improve the validity of research by providing a transparent and standardized way of accessing materials. The value created by the certification and distribution of biological materials arises from the very nature of the way biological research is conducted. Biological research depends on the development and execution of careful experiments that allow researchers to disentangle alternative hypotheses about the com-

provides a clear legal framework for domestic and international inventors to satisfy disclosure requirements. To receive protection, patented living organisms must be deposited at an international depository in the country in which protection is sought. Although protection is ensured during the life of the patent, the inclusion in the BRC depository results in researchers' independent access to materials after patents expire. Since the treaty's inception in 1981, fifty-four countries have ratified the treaty, and there are thirty-one authorized patent depositories around the world (including two in the United States, the ATCC and the Agricultural Research Service of the Department of Agriculture).

Although the function of BRCs as patent depositories was unclear initially, they have become important sources of protection for intellectual property while simultaneously facilitating increased exchange of patented materials. National BRCs have gained experience and established rigorous procedures to ensure that patent rights are upheld, and they have developed sophisticated materials transfer agreements that encourage licensing of intellectual property. These agreements have been particularly useful in enabling low-cost access to patented materials by university, nonprofit, and academic medical center researchers. Many companies have found that allowing researchers access to their patented materials has uncovered new uses and properties that have increased the commercial returns earned on the initial discovery. Overall, the Budapest Treaty has eased potential tensions between public and private research initiatives by fostering greater knowledge and materials exchange.

position and functioning of living organisms. In many cases the key to effective experimental design lies in understanding minute details about the properties of an organism, which allows alternative effects and mechanisms to be ruled out. By using biological materials whose properties have been characterized by previous research and that can be accessed through a BRC, scientists can dramatically reduce the experimental uncertainty associated with the scientific tests themselves. BRCs increase the integrity of the scientific process by providing independent access to a wide variety of standardized biological materials.

Knowledge Hubs: An Innovation Policy Framework

Although few doubt the value of science and technology in promoting economic growth, there is less consensus about how to design and implement effective policy initiatives. In many cases, even well-intentioned proposals contain neither clear statements of priorities nor guidance about how to choose among alternatives. Many policy challenges—from effective intellectual property rules to the implementation of the Convention on Biological Diversity—involve multiple elements, making policy analysis difficult.

The knowledge hubs framework helps policymakers develop and evaluate proposals that relate to control of and access to biomaterials. As knowledge hubs, investments in BRCs are complementary to publicly funded research investments and help maintain a high level of research productivity over time. However, in the current environment too few resources are devoted to ensuring that the results of publicly funded research are retained in institutions that contribute to step-by-step progress. While many analyses implicitly assume that cumulativeness is ensured by scientific norms or commercial motives, this framework highlights the crucial functions of nearly invisible institutions in facilitating this process. Effective policy should ensure and promote cumulative knowledge. This perspective has implications for a range of policy issues in biological research, including the level and allocation of funds, rules concerning scientific research and intellectual property disclosure, and the nature and optimal scope of activities that should be undertaken by particular knowledge hubs.

By highlighting the central role of institutions in ensuring cumulativeness (and the potential incentive problems facing individual researchers), the knowledge hubs framework offers policy guidance that accounts for the essential elements of the life sciences research environment. If knowledge hub institutions such as BRCs do not receive adequate funds and adequate involvement from the research community, there will be insufficient investment in the accumulation of knowledge. These insights motivate three crucial innovation policy questions:

What is the most effective mechanism for funding investment? By their very nature, knowledge hubs are public goods, and their funding and continued support may be subject to substantial free riding. Although the infrastructure for innovation in the life sciences is stronger when effective hubs are present, no company or research organization may have an incentive to maintain a hub (indeed, because providing transparency is a primary function of a hub, individual incentives may be particularly weak). This is not simply of theoretical concern; inattention and lack of continued support have hindered the development of knowledge hubs for more than a century. Private firms may have some incentive to supply some hub features, but private objectives will be at odds with the overall activities of an effective hub. While the best mechanism will depend on the particular issue, effective funding will likely involve collective action (for example, by the leadership of scientific societies or by a consortium) as well as direct government support.

What is the role of the research community in contributing to and accessing the knowledge hub? Even when funding is adequate, an effective hub relies on a cycle of deposit and acquisition of knowledge by members of the research community. In general the incentives for deposit may be limited because retaining control of knowledge may offer greater individual benefits. The development of institutions and procedures that ensure that credit (or royalties) are received by depositors increases the incentives for deposit and thus contributes to the sustainability of a knowledge hub such as a BRC. In some cases, it may be worthwhile to mandate deposit because although all members of the research community would be better off if everyone contributed, no individual member has a personal incentive to do so.

Can policy initiatives be implemented effectively? While knowledge hubs such as BRCs seem crucial in fostering cumulative progress, policy initiatives to encourage these institutions must not detract from the successful execution of their basic functions. Particularly in a changing environment, taking advantage of opportunities should be balanced with maintaining competence in core areas. Providing a public good whose value often is most apparent only after considerable time has passed, knowledge hubs must be careful to focus on those initiatives where they are likely to provide long-term value. Sometimes, maintaining focus may require establishing separate institutions to address different aspects of cumulative knowledge within a research community.

4

*A Penny for Your Quotes? The Impact of BRCs on Life Sciences Research**

Qualitative analysis suggests that the knowledge and materials associated with biological resource centers have an important impact on the cumulative research that is so central to scientific advancement, technological progress, and long-term economic prosperity. Chapter 3 identifies four distinct advantages associated with BRCs: authentication and certification of deposited materials, long-term preservation of materials, provision of independent access to materials, and exploitation of economies of scale and scope. BRCs amplify the eventual impact of a given piece of research, over a longer period of time and across a wider swath of the scientific community.

This chapter evaluates these claims regarding the impact of BRCs on the cumulativeness of research in the life sciences through a quantitative investigation of two features:

—citations (of either scientific publications or patents), which can serve as a useful though imperfect measure of the cumulative impact of discoveries.

—BRC deposits can be linked with specific scientific publications or patents, and the date of publication and the date of deposit may sometimes differ.

*Cowritten with Jeffrey L. Furman, Boston University School of Management.

This analysis exploits the fact that the process of knowledge accumulation in the life sciences is made "visible" through the citations associated with BRC deposits in the scientific literature. Indeed, over the past decade, researchers in economics and other social sciences have increasingly exploited citation patterns to gain insight into how scientific and technical knowledge is created and distributed.[1] Perhaps the key insight of this approach is that the number of times an article is cited (and by whom and how long after its publication) provides a useful (though noisy) index of the "impact" of that article on subsequent scientific research (see box 4-1).

We extend these prior analyses by linking scientific articles to specific BRC deposits. Our approach takes advantage of the fact that the leading biological resource center in the United States, the ATCC, collects bibliographical reference information for the majority of deposits in its collection (see box 4-2), including in its catalogue the name of the original depositor, the date of the deposit, and key scientific articles associated with the deposit. Moreover, in some cases, the date of the deposit is distinct from the date of initial scientific publication, offering a method for isolating the impact of deposit. We explore each of these features to develop data sets and assess key hypotheses about the impact of BRCs on cumulative research.

In particular, if depositing biological materials in BRCs increases future researchers' access to the knowledge associated with them (that is, increases its cumulativeness), then scientific articles associated with BRC deposits should be more intensively cited as a result of their greater impact on follow-on research. As knowledge hubs for life sciences research, BRCs amplify the impact of a given research article by increasing the level of confidence placed by researchers on its findings and reducing the costs of assessing materials to conduct follow-on studies. However, since articles that are associated with BRC deposits may simply be more influential pieces of scientific research, it is critical to identify a "control" group of articles of similar scientific "quality" and initial publication date that are not associated with a BRC deposit. It also is important for policy analysis to disentangle the intrinsic scientific importance of articles from the specific impact of a BRC deposit. If research is more likely to be cited (and over a longer period of time) when associated materials are deposited in a BRC, then BRCs exert an impact on the cumulativeness of life sciences research.

Our empirical assessment is divided into three parts. First, we evaluate the citations of research publications linked to BRC deposits and compare them with publications that are not linked to BRC deposits from the same journals published in the same time frame. The results are dramatic: BRC-linked arti-

---

1. Griliches (1998); Jaffe and Trajtenberg (2002).

*Box 4-1. The Use (and Misuse) of Scientific Paper
and Patent Citations*

The use of citation analysis to evaluate the impact of scientific research and technological innovation goes back to the pioneering work of Eugene Garfield in the 1950s (Garfield himself proposed and developed the *Science Citation Index*), and it has experienced an explosion since the early 1980s. Economists, sociologists, and policy analysts have increasingly exploited rich bibliometric data sets and sophisticated empirical tool kits to compile citation patterns in order to highlight key phenomena relating to scientific research and technological innovation. For example, for each scientific discipline there exists a "hierarchy" of journals to which researchers may submit papers for peer review and publication. Sociologists of science have documented the self-reinforcing nature of this publication hierarchy by showing that more prestigious journals, which tend to be associated with higher citation rates, attract higher-quality submissions. Economists, on the other hand, have exploited scientific and patent citation data in order to assess the importance of localized knowledge spillovers, intellectual property rights, and the impact of laws such as the Bayh-Dole Act (Griliches, 1984; Jaffe, Trajtenberg, and Henderson, 1993; Jaffe and Trajtenberg, 2002). Similarly, building on the pioneering database and analytical work of firms such as CHI Research, the National Science Foundation (NSF) has increasingly used citation analysis for assessment purposes. For example, in order to measure the relative scientific capabilities of nations, the NSF highlights the role of citations as "an indicator of the perceived influence of a nation's scientific outputs . . . and technical work" (National Science Board, 2002).

Along with this explosion in the use of citation data has come increased understanding of appropriate (and not so appropriate) uses of these statistics. Three methodological issues are central to the appropriate use of citation data. First, small differences in the citation rate of a single paper, particularly

cles had more than a 200 percent greater likelihood of being cited, a gap that became larger with the length of time since initial publication. That difference in likelihood of being cited was evident across several different collections of the ATCC, and it increased during the 1990s. The lead authors of BRC-linked articles tended to be located in the United States (not surprisingly, given that the sample was drawn from ATCC deposits), but they were otherwise similar to lead authors of the control articles. A smaller, more preliminary study of patents that draw on BRC materials reinforces the claim that BRCs are playing an increasing role in biotechnology innovation.

early in its publication history, are of limited value in determining the importance of the research or its use by the research community. Simply put, the publication process is subject to delay and errors in attribution, and so citation data tend to be noisy for a single paper or small group of papers. Citation measures are much more suited to comparing the overall impact of two groups of papers with similar ex ante characteristics. For example, in our present analysis, we compare citations to papers that were published in the same issue of the same journal but that differed in whether the materials identified in the articles had been deposited within a BRC. Second, it is important to distinguish between the legal interpretations of paper and patent citations. Paper citations usually recognize the perceived key influences on their authors' work but have no direct impact on the value of or right to the work, whereas patent citations have a legal status that "bounds" the new invention, since one cannot receive intellectual property rights that have already been granted to others. This analysis focuses mostly on citations of scientific research papers. Finally, nearly all citation databases are highly skewed; a very small number of articles (or patents) receive a very high share of the citations. This fact suggests that empirical findings should be checked to ensure that they are not being driven by a small number of outliers and that only empirical procedures that are appropriate for such skewed data should be followed. As we discuss in box 4-5 and appendix A, our use of "negative binomial count data" models allows us to account for these features of paper citation data.

Overall, the approach pursued here is based on the idea that paper and patent citations provide a useful though quite noisy estimate of the impact of the knowledge described within these publications. While the full range of citation analysis issues (and the extensive literature associated with these issues) is much too large to be summarized here, our analysis aims to adopt a state-of-the-art empirical approach incorporating key insights from this literature (Griliches, 1990; Jaffe and Trajtenberg, 2002; Adams and Griliches, 1996; and Moed and others, 1985).

The second stage of the analysis focuses on estimating the "marginal" impact of BRCs. By comparing how an article's citation pattern changes *after* materials associated with the article are deposited in a BRC, the specific role played by BRCs in amplifying the impact of research is isolated.[2] The results indicate that articles associated with BRC materials experienced a greater than 80 percent boost in their citations. Moreover, BRC deposit increased

---

2. As discussed in some more detail below, we use a "difference-in-differences" estimator, taking advantage of features of how deposits are made for specific collections.

Box 4-2. Choosing the Sample

The data in our analysis consist of two subsamples. For our "base" data set, we assembled a collection of research articles linked to BRC deposits, along with "matched" samples of control articles. To build the sample of BRC-linked articles, we took advantage of the fact that ATCC prepares reference information for materials deposited within its collections. For each ATCC deposit, the ATCC catalog identifies the deposit, the name of the original depositor, the date of the deposit, and key scientific articles associated with the deposit.* We selected 183 deposits at random from the materials deposited in three of ATCC's primary collections (which deal with bacteria, cell biology, and molecular biology) between 1984 and 1999, and we considered the first article listed within the ATCC deposit reference section as the "focal" article for each deposit. Multiple members of the scientific and information technology staff at ATCC with whom we conducted interviews suggested that the first reference article typically is the one most closely associated with initial use of the biological material.

For our more nuanced difference-in-differences analysis, we employed a second subsample, the special collections data, which was composed of 127 articles linked to materials in three special collections that had been transferred in bulk to the ATCC from private culture collections.** While the first subsample allowed us to evaluate a broad cross-section of BRC-linked articles, the special collections subsample allowed us to disentangle the intrinsic impact of the quality of the research from the impact of being deposited in a BRC.

Each of the ATCC-linked articles was matched with two types of control articles. Each type of control article was designed to be as similar to the ATCC-associated article on as many observable dimensions as possible. In the first set of controls, the nearest-neighbor controls, we used the National Library of Medicine's PubMed database to match each BRC-linked article

---

the "longevity" of the research impact—BRC-linked articles continued to be cited at a substantially higher rate, even twenty years after materials were deposited. The boost from deposit in a BRC also became far more pronounced from the mid-1980s. BRC deposit affected not only the number but also the type of citations; citations to articles after materials were deposited in a BRC tended to have higher citation rates themselves. Overall, as an economic institution, BRCs have played an important and unique role in promoting step-by-step progress in the life sciences.

Third, a rate of return analysis is presented, assessing how expenditures

with the article that immediately precedes it in the same issue of the same journal. For example, if an ATCC-associated reference was the third article in the June 14, 1986, issue of *Journal of Cell Biology*, the nearest-neighbor control article would be the second article in that same issue (in the event that the ATCC-associated article was the lead article in its particular issue, we used the second article in that journal as the control). This ensured that both the BRC-linked and control article had undergone the same review process and been published at the same time. Because some journals cover a wide range of scientific disciplines—for example, *Science* and *Nature*—we also collected citation data for another type of control. Using a National Library of Medicine algorithm, we identified for each BRC-linked article the "most related" article that appears in the same volume of the same journal. For example, if an ATCC-associated article was published in the June 1986 issue of *Cell*, the most-related article control would be the article in the 1986 volume of *Cell* whose research topic and themes were judged by the National Library of Medicine (based on keywords, title, and abstract) to be most similar to the ATCC-associated article.

We then used the *Science Citation Index* (as well as other bibliometric information) to assess the relative impact of these articles on subsequent scientific research. In particular, as described in the main text, we compared the citation patterns of articles associated with ATCC deposits to the citation patterns of control articles. Thus we were able to identify the impact of BRC material deposits.

*The catalog is maintained online at www.ATCC.org and historically was published in catalog form (a partial printed catalog is still available, but the dominant mode for accessing this information today is through the Internet catalog).

**Numerous scientists, research institutions, and corporations maintain private collections. With the exception of those collections operated by firms, many of these allow open access to their collections; on balance, however, they are less engaged in characterization, and knowledge of the contents of their collections is less well diffused.

on BRC deposits compare with expenditures on alternative research investments. Expenditures aimed at increasing the impact of published research were compared with expenditures aimed at encouraging new research. The results are dramatic. While the average "cost" per citation for university-based biological research is estimated at $2,400, the cost per citation for deposit in a BRC was less than $900, under the most expensive calculation. Relative to traditional grant activities, investments in BRC authentication, storage, and access activities dramatically improved the accessibility of knowledge and so facilitated cumulative knowledge production.

Table 4-1. *Citation Rates, BRC-Linked Articles Compared with Controls*

	Backward citations, total	Forward citations	
		Total through 2002	Average per year
ATCC sample, average	32.9	151.0	9.2
Nearest-neighbor controls, average	26.4	49.5	3.0
Most-related controls, average	29.7	79.2	4.7
Percent difference between ATCC and			
Nearest-neighbor	125	305	305
Most-related	111	191	196

Citation Patterns of BRC-Linked Articles

Our initial analysis compares the rate of citation for articles linked to ATCC deposits with the rate of citation for control articles judged to be of similar scientific quality. We constructed article sets, matching each BRC-linked article with a "nearest neighbor" article published in the same issue of the same journal as well as a "most related" article published in the same volume of the same journal (see boxes 4-2 and 4-3). Patterns of citations for BRC-affiliated articles and control articles, both of which had undergone the same scientific review process and been judged to have roughly similar scientific merit, provide an indication of their relative impact on subsequent research. Because each pair came from the same issue or volume of a journal, the opportunity to be cited by future researchers was virtually identical within each set.[3] Several striking findings emerged:

—*BRC-linked articles had a much higher rate of citation than articles published in the same volume of the same scientific journal.* While both the BRC-linked articles and the controls had comparable numbers of "backward citations" (that is, they cite a similar number of references), BRC-linked articles had a "forward citation" rate nearly triple that of control articles published on the preceding pages (nearest-neighbor article controls) and nearly double that of the most similar articles that appeared in the same volume (most-related article controls) (table 4-1). By 2001, BRC-linked articles received nearly ninety more forward citations on average than either type of control article (figure 4-1).

3. BRC-linked articles may be published a few months earlier or later in the volume than "most-related" article controls. Unless there is some reason to believe that BRC-linked articles are published earlier in the year than "most-related" control articles, this would not bias the results. Note that there appears to be no reason to believe that BRC-linked articles are published earlier or later in the year than "most related" control articles.

Box 4-3. Citation Analysis Data and Measures

The empirical analysis was based on a data set composed of BRC-linked articles and accompanying control articles. For each of these articles, we measured citations for each year from the year of publication through December 2001. Citation activity was measured as "forward citations," the annual number of citations received by an article, and "cumulative citations," the number of citations received through the end of 2001. Overall, the citation rate of articles in the sample was quite high compared with a completely random sample of life sciences publications. The high citation rate is not surprising, since articles associated with BRC deposits tend to be published in top-tier scientific journals such as *Science, Nature,* and *Cell.* As with other citation data sets, the distribution is quite skewed: many articles had no forward citations and a small number of outliers had more than a hundred citations in a single year.

The analysis includes additional measures for the timing of publication ("publication year"), the time of publication ("deposit year"), and the length of time elapsed since initial deposit ("age"). In some cases, the deposit year measure yields an error of up to a few months. As materials deposited with ATCC must undergo authentication and cataloging before they are available for public use, there is some delay between the announcement of a transfer and ATCC's ability to ship materials for scientific use. In some occasions, materials may be available for a few months before their accession is officially declared in a catalog or other ATCC publication.

While the date of publication varies from 1970 to 2001, all accession dates in this analysis occurred from 1981 onward (1981 was the first year of the articles we drew for the random sample and the first accession date for articles associated with the special collections). We also investigated the impact of several additional article characteristics, including the number of pages for each article, the number of authors, and the number of backward citations. We also collected characteristics of the lead author of each article, including institutional affiliation, whether the affiliated institution was a company, university, or government organization, and whether its location was foreign. Finally, for articles linked to BRC deposits, we collected the price charged by the ATCC for access, which averaged approximately $230 over the sample.

Figure 4-1. *Cumulative Citations, BRC-Linked Articles Compared
with Nearest-Neighbor and Most-Related Article Controls, 2001*

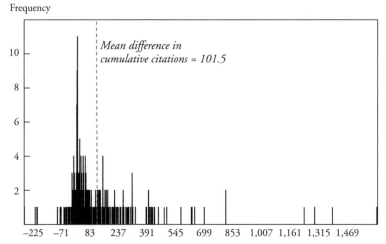

(Cumulative citations to BRC-linked articles) – (Cumulative citations to nearest-neighbor controls)

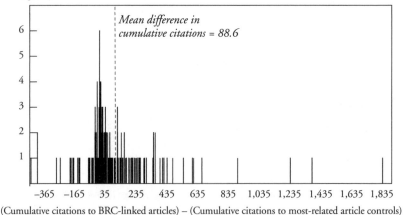

(Cumulative citations to BRC-linked articles) – (Cumulative citations to most-related article controls)

*—This citation gap appeared across article samples linked to several distinct
ATCC collections.* Figure 4-2 divides up the article pairs according to whether
articles were linked to deposits in the cell biology, bacteriology, or molecular
biology collections of the ATCC and compares citations for BRC-linked articles with those of the nearest-neighbor controls. The advantage experienced
by ATCC-linked articles appears for all collections, with the articles in the
cell biology collection experiencing the highest overall boost.

Figure 4-2. *Mean Difference in Cumulative Citations between BRC-Linked Articles and Nearest-Neighbor Controls, by Collection*[a]

Mean difference

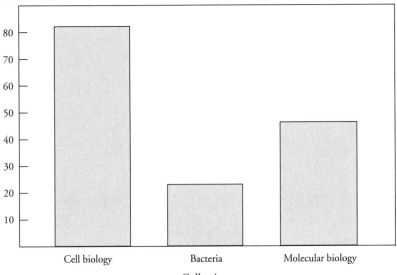

Collection

a. Based on a subset of the original data set for which the collection is identified.

—*The divergence in citations between BRC-linked articles and the controls grew with time since publication.* Perhaps no comparison exemplifies better how ATCC-linked articles differed from control articles than figure 4-3, which charts the articles' forward citation rates as their dates of publication became more distant. In the first few years after publication, BRC-linked articles received substantially more citations than did control articles. As time passed, all sets of articles received fewer citations; however, the gap between citations received by BRC-linked and by control articles increased substantially—reaching at least 250 percent ten years after publication. Even twenty years after publication, BRC-linked articles continued to average significant numbers of citations, while control articles received only one or two. In other words, although each set of matched articles was judged at the time of publication to have similar scientific merit, the knowledge described in BRC-linked articles tended to have a more pervasive impact on the research process, for a much longer period of time.

—*ATCC-linked articles and the controls had similar characteristics, except that lead authors were more likely to be located in the United States. Even after controlling for the differences in author and article characteristics, the citation gap between BRC-linked and control articles remained high.* As part of our

Figure 4-3. *Average Annual Citations by Age and Percent Difference, BRC-Linked Articles Compared with Controls*

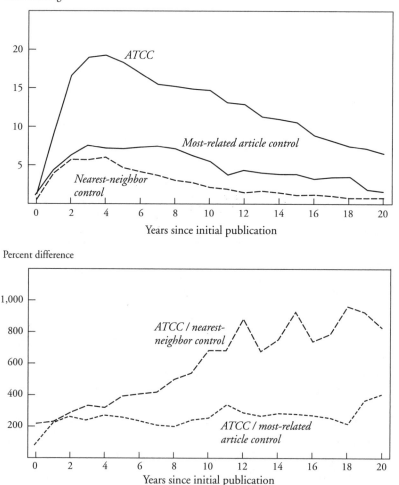

Annual average

Most-related article control

ATCC

Nearest-neighbor control

Years since initial publication

Percent difference

ATCC / nearest-neighbor control

ATCC / most-related article control

Years since initial publication

examination of the data, we undertook a number of tests to investigate how BRC-linked articles differed from the controls. Overall, the two samples seemed similar, with the exception that control articles tended to be published by non-U.S. authors and that one control set averaged somewhat fewer pages and more coauthors (table 4-2). Even after accounting for detailed characteristics such as these, the citation advantage experienced by BRC-linked articles remained similarly high.

—*An increasing number of patents acknowledge BRCs, such as ATCC, in*

Table 4-2. *Article Characteristics, BRC-Linked Articles Compared with Controls*

Characteristic	BRC-linked	Nearest-neighbor	Most-related
Number of pages	8.1	6.8	4.3
Number of authors	4.2	3.7	7.7
Type of author (percent)			
University	57.5	60.3	60.2
Government	14.7	13.1	11.8
Foreign	25.3	40.7	39.1

their scientific citations, abstracts, and descriptions. These BRC-referencing patents appear similar to other patents in terms of commercial and technological significance, although they do make more references to academic articles. To investigate the role that BRCs seem to have increasingly played in biotechnology, we also assessed whether ATCC reference materials contributed to the development of patented life sciences technologies. In particular, we tracked citations to ATCC materials in four key patent classes in which biotechnology innovation was most prevalent (figure 4-4). The results demonstrate a dramatic increase in the fraction of patents that referenced ATCC materials between 1981–2002. Taking a more in-depth look at eighty of these patents (drawn evenly from the four classes), we observe that patents that acknowledged their use of ATCC materials were more likely to reference nonpatent materials (like academic articles) in their applications. While such patents made more intensive use of science to develop their patented technology, they referred to prior patents and were referred to by future patents in numbers similar to those of a control group of patents issued in the same technology classes at similar times (figure 4-5).

The Effect of Deposit in a BRC on the Use of Scientific Knowledge

The analysis points to a dramatic difference in the forward citation patterns of articles linked to BRC deposits, even compared with articles that met the same publication standards. However, the approach so far did not isolate the "marginal" impact of BRCs in fostering cumulative knowledge production. It was possible that BRC-linked articles were of higher scientific importance than the controls we compared them with, notwithstanding their publication in the same issue of the same journal. To disentangle the effects of deposit and scientific importance, either of which might have led BRC-linked articles to be more highly cited, we turned to a more nuanced and revealing

Figure 4-4. *Percent of Patents Referencing ATCC, by Class and Period*

Percent

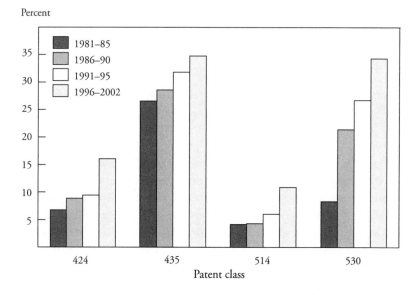

Patent class

Figure 4-5. *Characteristics, ATCC-Referencing Patents Compared with Controls*

Number of references

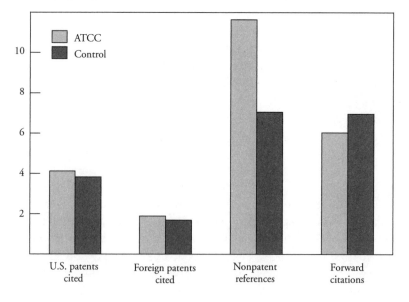

empirical methodology to determine whether the pattern of citation to a scientific article *changed* after the materials associated with the article were deposited in a BRC. We could infer from any changes in the citation pattern the direct impact that BRCs had on the use of this knowledge.

That approach took advantage of the fact that over the past twenty years a number of "orphan" collections have been "adopted" by large national BRCs. More specifically, some collections maintained in a specific university laboratory or research institute suffered a funding crisis or the retirement of key personnel, which led to a search for a new home (see box 4-4). These materials (often more than 100 distinct cultures) were transferred from the peer-to-peer network to a knowledge hub, often years after the initial publication of articles associated with their identification and characterization. This set of circumstances—in which materials were located for a period of time within the peer-to-peer network and then became accessible through a knowledge hub—allowed us to assess the specific impact that deposit in a BRC had on the use of this knowledge (see box 4-5 and appendix A). Several findings from this analysis stand out:

—*Depositing materials in a BRC resulted in a substantial boost in the citation of articles associated with the deposit.* To estimate the marginal impact of BRC deposit on yearly citations, a regression procedure was employed to account for the amount of time since an article's initial publication, the years in which citations occurred, and the overall impact of the article (independent of whether associated materials had yet been deposited in a BRC). This citation analysis measured two key values:

—the degree to which BRC-linked articles simply represented more important scientific research (selection effect)

—the degree to which deposit of materials in a BRC increased the use of the knowledge produced in the articles by members of the scientific community (impact of BRC deposit)

The results are quite dramatic. While the selection effect is associated with a 120 percent higher citation rate, the estimated marginal impact of BRC deposit increased the citation rate by 102 percent. In other words, when materials were deposited in a BRC, articles associated with those materials experienced a doubling of their citation rate relative to the pattern that would have been expected without the deposit.

—*The divergence in citations resulting from a BRC deposit grew over time after the deposit occurred.* The overall advantage resulting from a BRC deposit was not immediate; it grew gradually over a decade. Figure 4-6 illustrates that in the decade before deposit, articles associated with the special collections did not experience an elevated citation rate. However, within

---

Box 4-4. Special Collections

The second stage of the empirical analysis took advantage of the fact that "orphan" collections are occasionally transferred in bulk from non-BRC private collections. Such transfers usually result from a funding crisis or the retirement of key personnel at the institution in which the collection was stored. Of the special collections maintained by the ATCC, three were particularly interesting, since their inclusion in the ATCC was not directly related to an increased perception of their scientific importance.

The first is a set of articles associated with the Gazdar collection. This collection was transferred to the ATCC when Adi Gazdar left his position as head of the tumor cell biology section at the National Cancer Institute, along with his collaborator, John Minna, to become professor of pathology at the Hamon Center for Therapeutic Oncology at University of Texas Southwestern Medical Center at Dallas. The Gazdar collection was incorporated into the ATCC over a number of years; the materials examined in this analysis were accessioned beginning in 1994. The second set of materials was drawn from the Tumor Immunology Bank (TIB), which was transferred from the Salk Institute in 1981 because of funding considerations and was accessioned beginning in 1982. The final set of articles in the data set was associated with materials in the Human Tumor Bank (HTB). The HTB had been maintained by researchers at Sloan-Kettering until funding considerations led to its being transferred into ATCC beginning in 1981.

Historical details on ATCC's collections are drawn from discussions with Robert Hay, director of the Department of Cell Biology at ATCC.

---

four years of deposit, those articles experienced nearly a 50 percent boost, which grew to approximately 120 percent within ten years. Consistent with the key role that preservation and storage play in an effective knowledge hub, BRCs amplified the impact of scientific research findings over the long term.

—*The citation gap associated with deposit in a BRC has increased markedly since the mid-1980s.* We earlier observed that BRC-linked articles increasingly experienced a gain in their citation rates, starting during the mid-1980s. Figure 4-7 demonstrates this finding, focusing on how the marginal impact of BRCs changed over time. The overall trend in the impact of ATCC deposit was upward over time, and the 1990s witnessed a stronger impact than the 1980s. This recent increase in impact highlights the central importance of

Box 4-5. The Research Methodology

Two key methodological challenges were associated with isolating the distinct impact that BRC deposit had on forward citations. First, it is difficult to disentangle the marginal impact of BRC deposit from the possibility that materials ultimately deposited in BRCs are associated with articles of higher intrinsic scientific importance. We addressed this challenge by implementing a difference-in-differences empirical approach in which the estimate of the impact of BRCs was inferred from the *change* in citation patterns after BRC deposit (relative to change in the absence of deposit) (Angrist, Imbens, and Rubin, 1996). By simultaneously comparing citation patterns *across article pairs* (that is, comparing articles whose materials eventually were deposited in BRCs with those whose materials were not) and *across deposit-status within article* (that is, whether a particular article had yet been deposited), we could separately identify the degree to which articles had higher intrinsic importance (that is, the selection effect) and the degree to which ATCC deposit had a marginal impact on citation rates. Second, because citation data were realized in the form of annual "count" data and were highly skewed to the right (that is, the median was substantially smaller than the mean), the use of traditional regression techniques, such as ordinary least squares (OLS), was inappropriate. Instead, building on a decade of research in this area, we employed a negative binomial regression model, which accounted for both count data and the skewed citation patterns (see appendix A; Cameron and Triverdi, 1998; and Furman and Stern, 2004, for further discussion).

certified biological materials in an era in which bioinformatics and genomics play an increasingly important role in life sciences research.

—*The boost from inclusion in a BRC was apparent across a number of distinct special collections.* The examination drew from three distinct "orphan" collections transferred to the ATCC. Each of the collections that had been housed at ATCC for more than a decade experienced a substantial advantage in the citation rate of associated articles. While the Gazdar collection, which had been housed within ATCC for only a few years, has yet to experience a substantial overall boost, the Human Tumor Bank and Tumor Immunology Bank collections obtained citation boosts estimated to be 56 percent and 105 percent respectively.

—*Articles that cite articles associated with BRC deposits tended to be highly cited themselves.* Not only did BRC-linked articles have higher rates of cita-

Figure 4-6. *Number of Years before and after BRC Deposit*

Percent impact

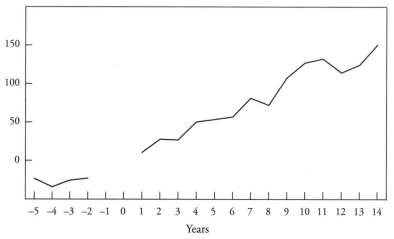

Years

tion than control articles, but so too did the "second generation" articles that cite BRC-linked research. Specifically, articles that cite BRC-linked research have approximately a 75 percent higher citation rate than articles that cite the control articles. This result is consistent with the view of BRCs as knowledge hubs that promote the diffusion of leading-edge scientific research. An analysis of citation patterns shows that articles that cite BRC-linked articles were otherwise similar to articles that cite nearest-neighbor controls, with the exception that a somewhat higher fraction of articles that cite BRC-linked research came from U.S. authors and that those articles had later average publication dates. The latter fact suggests that BRC-associated research streams enjoyed greater longevity.

The Cost Effectiveness of BRCs as Cumulative Knowledge Hubs

The results so far provide the first quantitative evidence of the role that BRCs play as knowledge hubs in the life sciences. Articles linked to materials deposited at the ATCC had a much higher rate of citation and that boost can be specifically attributed to BRC deposit rather than simply being associated with intrinsically more important research. However, these findings cannot provide effective policy guidance unless we first make at least a rough assessment of whether the benefits of making a BRC deposit are worth the costs incurred. A comprehensive cost-benefit analysis is beyond the scope of our analysis since we cannot fully capture the degree to which access to BRC

Figure 4-7. *Impact of BRC Deposit on Forward Citations,*
Marginal Effects by Year

materials improved the productivity of research. However, it is possible to offer a back-of-the-envelope estimate of the cost effectiveness of BRC deposits relative to public investments in traditional grant activities. In addition, although our analysis reveals the impact of BRC deposits at current collection levels, the benefits of deposits might ultimately run into diminishing returns; this analysis of the cost-effectiveness of BRCs is therefore most revealing about the rate of return of incremental expenditures, but it may not capture the dynamics of large-scale shifts in the size and scope of BRC collections.

The ultimate social value of publicly funded research is closely linked to the exploitation of that research by future researchers. Moreover, within a particular discipline, citations to an article offer a noisy but useful proxy of the degree to which this cumulative process occurs. While caution is important in interpreting citation-based results, the framework suggests that systematic increases in the rate of citation for a given type of article are evidence of a higher rate of exploitation of that knowledge by the research community.[4]

This insight is at the heart of our cost-benefit calculation. We compare how a given level of expenditures on BRC deposits compares with alternative

4. As emphasized earlier, we are not discussing the citation rate of an individual paper, but the average citation rates associated with classes of papers distinguished according to specific criteria and sampling.

research investments in terms of promoting cumulative progress. In other words, how do investments in BRC deposits and authentication activities compare with traditional grant programs in terms of their efficiency in seeding the knowledge stock of future researchers? This exercise involved the calculation of three estimates:

—*Baseline citation cost.* This is the cost per citation paid by public funding agencies (such as NIH) when allocating resources that result in published scientific articles. This cost is calculated using the estimates in a recent article by Adams and Griliches.[5] Using data from the 1980s, these authors estimated the relationship between expenditures and academic research output (papers and citations) for individual academic departments at top universities across the United States, including biology departments. Using these measures (and converting all expenditures into 1987 current dollars), they estimated the cost per citation to be $2,400 at a top-ten biology department and $4,200 at nonelite public universities. Using the BEA R&D price deflator to recompute this number in current dollars, we find the *lowest* Adams and Griliches estimates of cost per citation to be $2,887. Being conservative in estimating the effectiveness of BRC expenditures, we chose the lowest estimated cost per citation among the Adams and Griliches figures to set the baseline citation cost at $2,400 for the life sciences.

—*BRC accession cost.* This is the full cost of deposit and accession into a national BRC collection such as the ATCC. According to a recent OECD Report on Biological Resource Centers, the highest estimate of BRC accession cost was $10,000.[6] While it is likely that the true marginal accession cost may be somewhat lower than $10,000, we used this high-end figure to bias us away from finding evidence for the cost effectiveness of BRCs.

—*BRC citation boost.* This is the incremental number of citations expected to result from deposit and accession into a national BRC. We computed three different estimates of the BRC citation boost. The first two of these computations build on the data provided by Adams and Griliches.[7] In their work, the average biology article received 24.6 citations during the first five years of publication if authors were located at a top-ten university and 14.3 citations if authors were located at universities below the top ten (in biology). In our most conservative estimate, BRC deposit was associated with an 82 percent increase in citations. If we assumed that the marginal accessioned material came from a top-ten university laboratory, then the marginal impact

5. Adams and Griliches (1996).
6. OECD (2001).
7. Adams and Griliches (1996).

Table 4-3. *BRC Deposit Cost-Effectiveness Analysis*

Estimated baseline citation cost [a]	BRC accession cost [b]	Article category	BRC citation boost	BRC citation cost	BRC cost-effectiveness index
$2,400	$10,000	Articles published by "top ten" biology department	20.17	$459.7	4.84
		Articles published by random biology department	11.73	$852.8	2.81
		Average article associated with BRC deposit	11.14	$898.7	2.67

a. Estimated baseline citation cost drawn from Adams and Griliches (1996).
b. Estimated BRC accession cost drawn from OECD (2001).

from deposit was estimated to be 20.2; if the accessioned material was drawn truly at random, we assigned a citation impact of 11.7, based on the citation rates of articles published by authors outside the top ten. We also computed the BRC citation boost directly from the estimates provided in the last section, focusing on the incremental boost realized by BRC-linked articles within the sample. Using this formulation, the BRC citation boost was 11.1; interestingly, BRC-linked articles within the sample had a BRC citation boost quite close to the estimated BRC citation boost for articles drawn from random biology departments.

Dividing the BRC citation boost by the BRC accession cost yields an estimate of the BRC citation cost, which we then compared with the baseline citation cost. The results of this calculation are presented in table 4-3. These estimates are dramatic. Even using the estimates that result in a conservative calculation, BRC deposit expenditures offered at minimum a 267 percent efficiency benefit in terms of citations. Of course, these calculations must be interpreted cautiously given the noisiness of citation data. However, given that the primary criterion for current public basic research expenditures at NIH is the likelihood that the research will have an important impact on the discipline associated with it, this analysis suggests that the impact of already funded and published research may be amplified cost effectively through the authentication, storage, and independent access features associated with knowledge hubs such as national biological resource centers.

5

Policy Challenges and Principles: Implications for BRC Stakeholders

The value of biological resource centers derives from their ability to amplify the impact of life sciences research; their ability to maintain their value depends on effective policy. Though valuable, biological resource centers are inherently fragile as scientific institutions. Scientists draw on materials and expertise provided by BRCs in conducting their research, but they take full credit for whatever discoveries result. For publicly funded research, the rewards of discovery may include scientific prestige and career advancement; for private companies, the rewards offer the opportunity to earn commercial returns. In all but a few circumstances, researchers do not explicitly acknowledge the role of BRCs in their research, nor do researchers share any of the financial returns from their discoveries. By their nature, BRCs are simultaneously valuable yet invisible.

The potential costs of ineffective policies toward BRCs are much greater than the failure of any single research project—they could retard the rate of scientific and technical progress across all areas of life sciences research. The benefits of funding depend on the cumulativeness of the knowledge production process. Because researchers build on knowledge and discoveries from the past, they can maintain a high level of research productivity. By authenticating results and maintaining the integrity of specimens—and so enabling researchers to build on the discoveries of others without having to reinvent the wheel—BRCs play a unique role in facilitating cumulativeness. Without effective BRCs, there would be an overall reduction in research productivity.

Effective public policy toward BRCs should be designed to enhance the ability of BRCs to support cumulative scientific and technical progress. Government funding of basic life sciences research should explicitly account for cumulativeness through the process of research funding, the life sciences research community must actively participate in nurturing the cumulative research process, and BRCs themselves must build a strong and consistent performance record to achieve their full potential as knowledge hubs for the life sciences. Achieving these broad objectives within each BRC stakeholder community requires a clear understanding of the key issues that impinge on each of the three functional activities associated with BRCs: acquisition, authentication, and access. Therefore, before turning to a policy analysis for each of the BRC stakeholder communities, I first discuss some of the key issues for each functional activity.

Acquisition

By offering independent long-term access to authenticated biological materials, BRCs provide a standardized system for distributing materials among both public and private researchers. However, the impact of BRCs on cumulativeness depends on the quality and scope of the materials found in their collections. For a BRC to function effectively as a knowledge hub, its collections must hold materials with the greatest potential for follow-on research and overall scientific impact.

Two major challenges must be overcome in order for BRC acquisition policies to be most effective: the public goods problem and capacity constraints. First, individual researchers and companies must be convinced to deposit their materials with BRCs and relinquish control of the materials for the benefit of the broader research community. Though depositing materials in a BRC may increase the overall scientific impact of research, it also may reduce the ability of the contributing researcher to extract concessions from potential follow-on researchers. Consequently, without special inducements, a BRC deposit is subject to the public goods problem: the incentives for individual researchers to deposit materials are substantially lower than the benefits to the research community. This problem is exacerbated by BRCs' long-standing policy of not paying for collections or deposits. Many individual researchers consequently choose to maintain private control of their materials, particularly for collections with the highest potential for scientific or commercial impact.

Second, the ability of BRCs to accept deposits for accession and preservation is constrained by the level of available capacity and resources. Lack of capacity or financial constraints limit the ability of BRCs to accept acquisition opportunities as they become available. This can reduce the cohesiveness and consistency of collections and discourage individual researchers from depositing materials in public collections.

Both deposit incentives and funding constraints have affected the evolution of BRC collections. Even national collections that include an extraordinary range of materials, such as the ATCC, represent only a modest share of all biomaterials used in research.[1] Further, though many BRCs have attempted to adopt a priority system for their deposit activities, most deposits still result from an individual researcher's preferences or biases, idiosyncratic institutional circumstances, or the potential for commercial returns. The recent debate over stem cell research highlights the impact of public policy on the availability of biomaterials, with its potential consequences for the rate of scientific advancement and even international competitiveness (see box 5-1). With uneven budgets across locations and over time (and despite the diligent efforts of BRC personnel), there is no assurance that existing biomaterials deposit programs have systematically acquired those materials yielding the greatest benefits to the scientific community.

The evolution of BRCs into international patent depositories under the Budapest Treaty is perhaps the best example of a policy outcome that confronts the public good and capacity problems of BRCs directly. Though initially considered a modest initiative, the establishment of national BRCs as patent depositories has become a key driver in the expansion of collections over the past twenty years. The establishment of international patent depositories enhances collections in two ways, both through the inclusion of patented materials that fall under the terms of the treaty and by expanding the financial resources available to national BRCs that allow them to undertake additional collection and authentication activities.

The overall impact of BRC collections behavior depends on the availability of adequate resources and facilities, participation by the scientific and commercial community, and effective operation by BRCs themselves.

---

1. While BRCs hold only a minority of all potential materials used for research, BRC collections likely hold a strong majority of those materials used in the greatest number of experiments. In another words, when use by researchers is considered, BRC collections represent a substantial share of all research resources.

Authentication

BRCs enhance research productivity by providing a centralized hub for accessing authenticated research materials. When individual researchers have a higher level of confidence in the fidelity of biomaterials, they can reduce their investment in replicating research and focus instead on extending the scientific and technological frontier. Misidentification and contamination of materials dramatically reduce research productivity. A single mistake calls follow-on research into question, and all researchers must invest more effort and time in duplicating prior results. Moreover, when adequate controls are not in place, there is no guarantee that individual researchers are making the effort necessary to replicate previous research. The scientific and technological fields place a premium on novelty and provide few if any explicit rewards for independently verifying research findings.

Ensuring the validity of experimental biological materials requires coordinated action on the part of all BRC stakeholders. Consider the outcome of the HeLa contamination scandal. After some initial delay, government agencies such as the National Institutes of Health provided substantial funding and grants for authentication research, the life sciences research community directed effort and attention toward the development of testing protocols, and national BRCs such as the ATCC and DSMZ were involved in both the development of validation technology and the diffusion of the resulting validation techniques.

Access

Finally, enabling cumulative progress through the sharing of research resources involves two steps: preserving research materials from one generation to the next and making them available on a fair and reasonable basis. Without effective preservation (degradation of materials limits their culturability), researchers will be able neither to confirm (or refute) prior findings in a precise manner nor to build on prior discoveries through the use of similar materials. Even with materials that have been adequately preserved, access depends on the innovator's willingness to share and ability to devise an effective distribution system for doing so.

Overall, an effective BRC access policy should attempt to provide trained researchers with authenticated materials at (or near) the marginal cost of distribution over the long term. This objective, however, must be balanced against the necessity of achieving at least three competing objectives:

Box 5-1. Stem Cells: A Balance between Science and Politics

Since the isolation of human stem cells by James Thomson at the University of Wisconsin in 1998, their use has been a source of both political and scientific controversy. Stem cells are unspecialized cells that can renew themselves for long periods through cell division; because they can be induced to become cells with specific functions, they have potential therapeutic applications (National Institutes of Health, 2002; Lee, 2003). However, a continuing ethical debate rages over the implications, moral and otherwise, of harvesting stem cells from human embryos. Opponents of the use of embryonic stem cells claim that life begins at the moment of conception, making the destruction of embryos to obtain stem cells morally reprehensible. On the other hand, advocates of embryonic stem cell research emphasize its therapeutic and scientific benefits, ranging from cell transplantation in the treatment of Parkinson's disease to organ replacement. Supporters also argue that concerns about morality are misplaced because embryonic stem cells can be taken in the earliest stages of development, prior to implantation of the embryo in the uterus.

Beyond moral concerns, the potential of stem cell research, which clearly is in its earliest stages, is undeniably vast. It is at the earliest stage of study in a new area that the key elements of the research infrastructure are developed and norms and rules regarding resource sharing are established. Since the science is in such an early stage and techniques to maintain and assess materials are still being developed, research productivity will be strongly influenced by whether scientists are able to access the widest variety of high-quality stem cell lines that could be made available.

Though elements of this infrastructure are being built in the United States, the Bush administration has placed extremely strict limitations on the use and development of materials, particularly in comparison with those of other nations. In particular, embryonic cloning procedures are banned in federally

—funding BRC acquisition, authentication, and preservation activities
—respecting the intellectual property claims of biomaterials developers
—protecting against biosecurity risks.

Since a range of current public and private policy issues hinge on these trade-offs, we consider each separately in turn.

Balancing Access with the Costs of Acquisition, Authentication, and Preservation

Consider the role of user fees in recouping the fixed costs associated with developing and maintaining BRC collections. On the one hand, researchers

funded research, and only sixty-four specific human embryonic cell lines may be used, all of which existed prior to the establishment of the policy in August 2001. To implement this policy, NCRR, in conjunction with ATCC, established the National Stem Cell Resource, which serves as a hub for nonhuman stem cell materials; NCRR also established specific resource centers to regulate access to the sixty-four "established" cell lines permitted under NIH policy.

However, other nations are establishing the required infrastructure for both embryonic and nonembryonic stem cell research, for both humans and nonhumans. For example, Australia established early leadership in stem cell research with the development of BresaGen, a start-up company with ownership rights to four key stem cell lines. Perhaps interestingly, BresaGen has chosen to freely license the use of its stem cell lines at marginal distribution cost; however, use of the BresaGen cell lines requires consenting to a royalty agreement if the research results in commercialized products (NIH, 2002).

The Australian national government also established the National Stem Cell Centre in Melbourne, and legislation allowing embryonic stem cell research passed after vigorous public debate. Together, these investments and policies have helped Australia establish the foundations of a strong research and commercialization infrastructure for stem cell research (Gans and Stern, 2003b). Other aggressive efforts to take advantage of the caution inherent in U.S. policies are under way in numerous countries, ranging from Sweden to China and even to Saudi Arabia (*USA Today*, 2002).

Though the United States has a preexisting advantage in personnel and the capabilities required to take advantage of breakthroughs in stem cell research, effective cumulative research and successful commercialization depend on the nation's research infrastructure. As a result, though stem cell research can (and still might) be an area of growth for American companies, the fruits of stem cell research are just as, if not more, likely to be exploited abroad first.

tend to access only a very small number of materials on a regular basis. For example, the fifty most popular materials may account for as much as 90 percent of total materials requests for a large BRC. On the other hand, as highlighted by examples such as *Thermus aquaticus* in the development of the polymerase chain reaction technique, neither BRC personnel nor individual researchers can accurately predict which materials are likely to have the greatest impact over the long term. As a result, comprehensive and well-characterized collections take decades to build and require the costly preservation of a very large number (potentially tens of thousands) of deposits.

Since most deposits are rarely requested, funding acquisition and preservation costs solely through access fees would require subsidization of low-use materials by high-use materials. The most popular biological materials would be distributed at well above their marginal distribution cost in order to fund the acquisition and preservation of less-used materials, raising the cost of access for most researchers and limiting the diffusion of biological materials. Perhaps fortunately, then, the need for subsidization is constrained by the presence of the for-profit biomaterials industry, which focuses on the small number of high-volume products. Cherry picking in the biomaterials industry is particularly easy: a private firm can simply order a culture sample from a BRC and then sell samples of that culture to both university and private researchers. As a result, private sector competition limits the ability to subsidize BRC collection and authentication activities. Overall, while different countries and regions fund BRC acquisition and preservation costs in various ways, effective policy must attempt to keep the costs of access low (near marginal distribution cost) while allowing well-run BRCs to raise sufficient revenues to help fund collection and preservation activities.

Balancing Access against the IP Claims of Biomaterials Developers

Intellectual property concerns also affect access to biological materials. The increased propensity for researchers to patent their discoveries may limit the research community's access to many research materials. BRCs balance IP rights against the need for access through materials transfer agreements (MTAs), which offer nonexclusive licensing rights to BRC users. By enhancing the effectiveness of the market for the exchange of licensed materials, BRCs have increasingly come to serve as key knowledge brokers for researchers throughout the life sciences.[2]

Balancing Access against Biosecurity Risks

Access policies must address biosecurity and bioterrorism concerns. Effective policy in this area must balance the need for oversight with the need for research, both of which have played a role in the aftermath of the anthrax attacks of October 2001. On the one hand, minimizing the likelihood of attacks (and increasing the effectiveness of investigations) requires comprehensive monitoring of the control and distribution of selected materials, and since October 2001, policy has been retooled to ensure that that is done (see box 5-2). Especially compared with the patchwork system developed during the 1990s, recent policy changes provide a firm foundation for limiting the

---

2. Cypess (2003c).

distribution of selected biomaterials. On the other hand, providing regulated access to these materials to a wide range of researchers remains critically important in confronting bioterrorism, if and when it occurs, through the development of detection systems, medical treatments, and more effective defenses. For example, recent improvements in detection technologies were achieved only by allowing the broader research community access to certain materials. Addressing the challenges of bioterrorism requires not only a robust program of government funding and investment, but also active involvement and responsibility by the scientific community and the development of effective programs and operations within BRCs themselves.

Overall, the knowledge hubs framework offers insight for a range of contemporary policy challenges across BRCs' three broad functional areas: acquisition of materials, authentication of materials, and provision of long-term access to materials. As summarized in appendix B, addressing these challenges requires action on the part of each BRC stakeholder community, including government, the life sciences research community, and BRCs themselves.

Implications for Government Funding Agencies

Since the clear call for federal investment in basic research by Vannevar Bush more than fifty years ago, public support for research has been premised on the fact that discovery and invention are, by and large, *public goods*. Market forces, by themselves, will underinvest in research activities whose benefits are widely distributed.[3] There will not only be underinvestment, but misallocation of funding also may occur. Specifically, while substantial public and private funds are now devoted to the *production* of research, there are few concrete mechanisms ensuring (or providing the necessary funding to ensure) the *cumulativeness* of research.

Though the achievements of BRCs often are rendered "invisible" by the very researchers who exploit them, BRCs serve a vital public interest and have demonstrated historically high rates of return in terms of the promotion of future research. To build on those returns, BRCs must have access to adequate resources to accomplish their mission and objectives. Whereas competitive funding, scientific prestige, and commercial opportunities have facilitated tremendous growth in aggregate support for life sciences research, funding for life sciences research infrastructure—and biological resource centers in particular—remains fragile and neglected. Effective investment and

---

3. Bush (1945).

Box 5-2. The Fight against Bioterrorism

The September 11, 2001, terrorist attacks left the scientific community—and the rest of the country—in a state of uncertainty over how to address a host of new security challenges. Concerns mounted over supervision of researchers (background checks, invasive security measures, and so forth), censorship, and immigration/travel obstacles for foreign scientists. International projects were approached cautiously as the nation responded to the war on terror.

The anthrax mailings that followed in October 2001 raised new concerns that directly involved the life sciences research community. The "anthrax mailer" possessed capabilities that led many to suspect that those responsible included a sophisticated scientist with access to anthrax organisms from specialized sources (Candiotti and Mears, 2001). The FBI and other agencies attempted to track the location and ownership of various strains of anthrax and other hazardous biological agents. By and large, analysts outside the biological resources community expressed sharp concern with the absence of registration requirements for biomaterials that could affect national security. Simply put, because those with anthrax had no duty to disclose their holdings, it was impossible to construct a "census" of the holdings of selected materials.

Various legislative initiatives in bioterrorism reflecting the security and research concerns of the scientific community appeared in 2001 and 2002. Presidents of the major scientific societies, including the National Academy of Sciences, the National Academy of Engineering, and the Institute of Medicine formed the Committee on Science and Technology for Countering Terrorism to advise Congress on how best to mobilize the U.S. scientific workforce to counter terrorism threats (National Academies, 2002). Over the past two years, academic and government scientists have been at the forefront of the efforts to ensure security, articulating new priorities and establishing new policy in an era in which science and technology are simultaneously the source of new terrorism threats and the best hope for countering them.

Two notable pieces of bioterrorism legislation, including provisions in the Patriot Act and the 2002 Public Health Security and Bioterrorism Prepared-

funding policies should therefore be guided by the "cumulative knowledge investment principle": government funding of basic life sciences research should take its effect on the cumulativeness of knowledge explicitly into account. The cumulativeness criterion should influence the types of projects funded, the terms on which results and materials developed by researchers are made accessible over the long run, and how grant performance and fulfillness and Response Act, have passed in the interim. Together, the new legislation revamps the rules and procedures governing access to certain materials. The Patriot Act requires those possessing "biological agents, toxins, or delivery systems" to register with the government and to provide "reasonable justification" for their possession of them, and it bans "restricted persons" from owning selected agents (Environmental Health and Radiation Safety, 2001). The Public Health Security Act assigns executive agency responsibility for enforcement and requires scientific organizations to report their current holdings of specified materials (U.S. House of Representatives, 2001).

These new regulations have resulted in the creation of procedures to determine the location and ownership of the country's current stock of regulated agents and the screening of personnel who work with them. Notably, sharp restrictions have been placed on the activities of foreign nationals, particularly from Middle Eastern countries such as Iraq and Iran. New security requirements governing the disposal of regulated materials also have been issued.

The battle over how to pay (and who should pay) for the cost of implementing the new regulations is just beginning. Though more than $1 billion was appropriated for bioterrorism defense in 2002, the allocation (and ultimate level) of these funds is still being debated. For example, current appropriations are insufficient to simultaneously upgrade Centers for Disease Control and Prevention facilities, improve the effectiveness of bioterrorism research, and fund state and local bioterrorism defense programs (American Association for the Advancement of Science, 2002a, 2002b).

BRCs have a critical role to play in this new environment. Over the past decade, as rules and procedures governing the transfer of selected materials have become more onerous, BRCs have increasingly become the de facto conduit for exchange of these materials. Indeed, there are few circumstances in which laboratory-to-laboratory exchange provides adequate documentation and security. Accordingly, CDC should specify a strong preference for the use of BRCs in complying with new requirements governing the distribution of select materials.

ment are measured.

Most policy analysis assumes that cumulativeness flows automatically from the scientific research process: as long as funded research is published in scientific journals, future researchers can build on prior discoveries. Instead, the history of the life sciences—from the problems of HeLa cell line contamination to the application of *Thermus aquaticus* in developing the polymerase chain reaction (PRC) technique—highlights the fallacy of this premise and the crucial importance of maintaining institutions that facilitate the cumulativeness of knowledge. BRCs cannot provide a "free lunch" for life sciences researchers. Cumulativeness requires significant investment in institutions that preserve certified biomaterials and allow materials to be shared by generations of researchers.

Acquisition Policies

Public policy toward BRCs should focus on establishing well-defined priorities for making acquisitions and ensuring that adequate funds are available. In particular, federal policy should encourage researchers and BRCs to undertake a more proactive approach to encouraging deposit and acquisition of biomaterials. Two specific new initiatives are particularly worthwhile:

—*Depositing biomaterials in a BRC should be treated as a preferred form of de facto compliance by researchers with federal grant requirements to provide independent and nondiscriminatory access to novel biomaterials.* Current requirements specify that federally funded research that results in novel discoveries—including most novel or unique biological materials—must be made available to other researchers on a timely basis. For NIH grants in excess of $500,000, a plan for providing access to resources used in making such discoveries must be submitted in conjunction with the research application. Adherence to the plan is a prerequisite to grant fulfillment and performance. Current policy affords researchers wide latitude in complying with access requirements and has few monitoring mechanisms. Depositing materials in a BRC yields lower costs of distribution and eliminates the potential for researchers to extract concessions from follow-on researchers. Establishing deposit in a BRC as the preferred form and de facto fulfillment of NIH and NSF policy permits a greater range of materials to be exchanged across a wider set of researchers at a lower cost.

—*Funding should be provided for biomaterial "buy-outs" to acquire high-priority resources protected by intellectual property rights.* While free and voluntary deposit of materials should remain the norm, stakeholders—including the federal government—should be willing to compensate depositors for materials with high potential for scientific impact that are protected by intellectual property rights. Acquiring intellectual property in addition to materials allows BRCs to diffuse key materials more effectively and lower the cost of acquiring materials for individual researchers. Buyout plans may be particularly important in areas where multiple competing IP claims are asserted over basic research "tools." A modest level of compensation to facilitate low-cost distribution of particular materials pays for itself in terms of higher research productivity and reduced expenditures on license fees.[4]

These initiatives will allow BRCs to undertake a proactive and systematic approach to acquisition, reduce inefficient transfer through the peer-to-peer network, and lower the long-term cost of employing standardized reference materials. Together, they promote increased potential for cumulativeness in life sciences research.

Authentication Policies

Public policy toward BRCs should emphasize the development, mainte-nance, and use of standardized and certified reference materials. A shift toward emphasizing the importance of certification requires a shift in the pri-orities for funding research projects, in researchers' responsibilities toward the wider research community, and in the level of support for developing and diffusing authenticated biological materials. Several specific initiatives can substantially enhance the fidelity of biological materials:

—*The criteria for funding basic research should explicitly emphasize research procedures that facilitate the cumulativeness of knowledge, including reliance on standardized reference materials.* Current NIH and NSF peer review criteria do not account for whether proposed research employs authenticated biolog-ical materials.[5] Researchers are neither rewarded for employing authenticated materials nor required to justify their use of noncertified materials in a given situation. While the use of noncertified materials is appropriate in many cases, researchers should be required to justify their decision to use them. While the intrinsic scientific merit of a project must always be considered paramount in awarding a grant, the effectiveness of NIH and NSF grant funding can be enhanced by encouraging researchers to employ research pro-tocols that build on prior research to avoid methodological problems such as contamination and misidentification of materials.

—*The terms of basic research grants should explicitly require that researchers provide nondiscriminatory access to authenticated biological materials and docu-*

4. Of course, it is important to prevent individual researchers from effectively "holding up" grant agencies funding a buyout. At least in part, this problem can be avoided by developing a standardized procedure for evaluating claims and making take-it-or-leave-it offers that are enough to induce patent holders whose principal source of revenue will be license fees to partic-ipate. This would lower the transaction cost of such agreements through the use of standardized materials transfer agreements (MTAs).

5. Under current NIH peer review guidelines, research grant proposals are "scored" on five distinct dimensions ("significance, approach, innovation, investigator, and environment"); together, these measures are supposed to encapsulate the novelty and feasibility of the research proposal.

mented data sets and research resources developed with grant funding. Whereas the current NIH and NSF peer review systems emphasize research novelty and feasibility, enabling future research should be an explicit funding criterion. Under current NIH and NSF peer review criteria, the ability of future researchers to exploit research findings is not directly considered in the review process.[6] For example, while recent NIH rules require that grant applications in excess of $500,000 include a plan for data access, these plans are not considered in the formal evaluation process. The contribution of research to cumulative progress is treated as an administrative rather than substantive aspect of research funding. Grant guidelines do not require researchers to authenticate their research findings through BRCs or to ensure the fidelity of their materials and data. Though such procedures are widely agreed to be good scientific practice, grant reviews do not evaluate them. NIH and NSF guidelines should include a grant's potential contribution to the cumulativeness of knowledge as an individual scoring criterion and incorporate the ability of researchers to offer authenticated research resources to future researchers as a substantive aspect of proposal review.

—*Investments in research infrastructure to enhance the fidelity of research resources should be substantially expanded over time.* The National Center for Research Resources (NCRR) and related agencies provide crucial support for the development of authenticated collections. However, the budgets of these agencies are quite modest (the total NCRR budget is well below 5 percent of the total NIH budget). This imbalance should be rectified. As the analysis in chapter 4 shows, the "rate of return"—in terms of the promotion of future research—to investments by such agencies in authenticating materials and facilitating exchange is greater than that achieved simply by expanding the number of NIH-funded research projects. Expanding BRCs' capacity to provide a wider range of authenticated biological materials and their ability to make systematic acquisitions should be an important priority and an area of renewed emphasis. In addition, federal agencies should fund and encourage research in the development of tools to enhance the quality of certification procedures. For example, though bioinformatics tools are still in their infancy, their potential to enhance the characterization of BRC holdings is immense; if successful, they could dramatically improve the link between biological

6. While the "significance" criterion does instruct reviewers to assess the effect of the research on "the concept or methods that drive this field," there is no explicit emphasis on whether the proposed research will enable others to exploit the proposed advance. Further, although the scoring system does reward the "development of new methodologies or technologies," it does not address whether such advances can be exploited at reasonable cost and with a high level of confidence by the wider research community.

information databases and the physical materials on which the information is based. More than a single project in isolation, infrastructure investments are crucial for the long-term productivity of life sciences research.

Access Policies

An effective BRC access policy must balance the benefits of providing researchers with authenticated materials at (or near) the marginal cost of distribution against the need to fund BRC acquisition, authentication, and preservation, respecting the IP claims of biomaterials developers, and ensuring biosecurity. With these concerns in mind, several policy changes would facilitate the provision of access to biological materials throughout the life sciences:

—*Federal funding for life sciences research must continue to require researchers to offer access to novel materials after a reasonable length of time.* The National Institutes of Health and the National Science Foundation should establish a proactive campaign to monitor compliance with access rules. While recent changes in NIH grant procedures require a plan for data access to be submitted with proposals that exceed $500,000, the lack of enforcement hinders effective biomaterials distribution.

—*The capacity to preserve biomaterials and preservation techniques should be substantially expanded over time.* Access to biomaterials depends on a stock of materials effectively preserved over time. While standard procedures in individual laboratories preserve materials for a few years, useful applications for many biological materials become apparent only decades after their initial discovery. Without long-term preservation, the research community cannot access "old" materials to advance "new" science (see box 5-3). Building on substantial progress in storage capacity and techniques by BRCs over the past two decades, NCRR and related agencies should encourage investment in capital equipment and research to enhance the overall storage capacity for biomaterials used in life sciences research.

—*Effective access to biomaterials should be facilitated by a competitive market for biomaterials that meet observable standards for quality and safety.* Though the pricing of specific materials should be at the discretion of individual collections, the socially optimal access cost is equal to the cost of distribution. Encouraging a competitive market for biomaterials, including the presence of for-profit culture distributors, offers the best hope for efficient pricing of biomaterials. However, it is important that this market be subject to strict quality and safety standards, since cumulative progress depends on the continued fidelity of biomaterials. Funding for acquisition, certification, and preservation activities should not rely exclusively on cross-subsidization of public BRC materials. Instead, the "fixed costs" associated with BRC activities

---

Box 5-3. Obolensk: Preserving the Soviet Collection

During the height of the cold war, the Soviet Union established an enormous life sciences research facility, the State Research Center for Applied Microbiology (SRCAM) in Obolensk, Russia. Though its stated purpose was to perform basic research in biology and medicine, the Obolensk complex was the centerpiece of the Soviet Union's enormous (and illegal) biological weapons program, Biopreparat. So secretive that it did not appear on Soviet maps, Obolensk at its height employed more than 3,000 research personnel, many of whom worked, in one way or another, on understanding the properties of biological organisms with weapons potential (*Wall Street Journal*, 2002). With more than 2,000 pathogens (including about 3,000 different strains of anthrax), Obolensk represented the largest and richest collection of such agents in the Soviet Union.

While the end of the cold war and demise of the Soviet Union brought an end to biological weapons research at Obolensk, the culture collection remained intact. However, economic woes within Russia led to sharp budget cuts, resulting in a 50 percent reduction in staff. During the late 1990s, Obolensk scientists went for more than eighteen months without pay (*Wall Street Journal*, 2002). Concerned that the security of the materials and scientific expertise at Obolensk would be compromised by its financial instability, the United States and other western governments contributed millions of dollars to form the International Science and Technology Center (ISTC) in order to maintain control of the collections and avoid the defection of scientists to emerging "opportunities" in Iraq and elsewhere. Over time, ISTC has spent between 25 and 33 percent of its $75 million on providing compensation to Soviet scientists, equivalent to between $25 to $35 a day (Ingram and Shargodsky, 2002). The dangers inherent in failing to provide adequate funding are highlighted by Victor Gusev, deputy general director at Obolensk, "They will most likely go to one of the so-called axis-of-evil countries. All they need is one month and they could make biological weapons for anybody" (Nuclear

---

should be met through a combination of new research grants (funded by NIH and related agencies) and from value-added services (such as acting as an international patent depository).

—*International patent depository requirements should facilitate offering nonexclusive and independent licenses to use patented biological materials.* At the time of the Budapest Convention, few would have predicted the pervasive impact that international patent depositories would have on the management

Threat Initiative, 2002). Besides providing incentives for scientists to stay, a key priority is to ensure the security and integrity of Obolensk's culture collection. Obolensk has a unique collection and contains key reference materials required for systematic biosecurity research. Following a 2001 agreement between President Bush and President Putin to cooperate on bioterrorism research, Obolensk scientists have been working with researchers from the United States and European Union. Among other initiatives, this collaboration includes a "strain exchange" program with CDC and joint research on materials with university and government researchers in the United States (*Wall Street Journal*, 2002).

While culture collections can take years or even decades to create, they can be destroyed overnight—or even faster. Despite the potential contribution of the Obolensk collections to antiterrorism efforts, the facility was faced with the threat of an electrical power outage in the spring of 2002. Obolensk managers had been unable to pay the electricity bill for fourteen months after a reduction in government spending on operations. Even a brief loss of power would have threatened the unique collection, resulting in vats of dead and damaged pathogens and reagents. As ISTC director Randall Lee Beatty remarked, "This is a crisis . . . this is one of the most important archives for dangerous pathogens in the world, and it would be a shame if it were destroyed for not paying the light bill" (Tyler, 2002). While the Russian government ultimately intervened with emergency funding, the status of the Obolensk facility is now under debate; under Russian law, the electric company can take possession of the facility for nonpayment of bills, without making any specific provisions to protect the collections (GBACI, 2002).

BRCs clearly play a key role in preserving biological materials. The history of Obolensk suggests that while most BRCs must survive in an uncertain (and changing) funding environment, the damage from inconsistent funding can be enormous. Not only does it threaten long-term preservation of existing collections, it also detracts from expansion activities.

of intellectual property and knowledge in the life sciences. Overall, the Budapest Treaty has facilitated the free exchange of biological materials, even those protected by patent. At the same time, the patent depository function has provided a more stable financial foundation for BRCs, particularly in the United States. While privately funded life sciences research has increased the share of research covered by patents, the use of standardized materials transfer agreements can enhance access to patented materials. The U.S. Patent and

Trademark Office should explicitly encourage the use of materials transfer agreements, particularly for materials used principally in research.

    —*As one of the primary distributors of regulated materials, biological resource centers should play a central role in ensuring biosecurity.* Most discussions of bioterrorism focus on a false trade-off between effectively promoting science and ensuring national security. In reality, the failings of the scientific community that have led to breakdowns in scientific and research integrity—the failure to adequately track, maintain, and authenticate research materials—are the same as those that pose significant security concerns. Both cumulative research and national security turn on the availability of a central database and repository to control access to and track the ownership of regulated materials, such as the population of anthrax strains. By certifying and standardizing materials, BRCs can systematically record their ownership, control, and distribution—the very knowledge required for an effective bioterrorism policy.

Overall, public policy toward BRCs should be based on the key role that they play in promoting cumulative scientific and technical progress. The NIH, NSF, and related agencies play a key role in shaping the environment for biomaterials exchange and in the sharing of research resources. The federal government therefore has a proactive role to play in enhancing the quality and scope of BRC collections, developing the technologies and tools available for authentication and preservation, and ensuring nondiscriminatory and independent access to the widest range of biological materials.

Implications for the Scientific and Commercial Research Communities

The effectiveness of BRCs in promoting cumulative knowledge production depends on the active participation of the life sciences research community. Decisions regarding individual materials exchange depend in good part on the policies and procedures adopted by the overall research community, including scientific societies, journals, and industry associations. The life sciences research community should be guided by a "cumulative knowledge participation principle": accepting principal responsibility for the continued and enhanced vitality of BRCs and related institutions. Indeed, while some areas of research have been marked by secrecy and inefficient duplication, some parts of the scientific community have proactively adopted norms and practices that encourage the sharing of biomaterials and research resources. More generally, the research rules, procedures, and investments of both scientific and commercial research organizations must take into account the long-term social benefits that result from each of the functional activities performed by BRCs.

Acquisition Policies

As key knowledge hubs in the life sciences, BRCs facilitate certification and independent access. While incentives for developing and characterizing novel biological materials must be maintained, scientific and industry organizations can leverage recent federal policy changes that emphasize the importance of disclosure and sharing of materials and resources. By adopting concrete rules and procedures that encourage the sharing of research resources, these organizations can help to provide a sustainable foundation for research productivity throughout the life sciences. Policies to achieve this objective include the following:

—*Journals should require authors to offer access to their research materials after publication of their research. Enforcement of access policies should be proactive, and depositing materials in a BRC should be treated as de facto compliance.* Publication in scientific journals is the principal way in which new discoveries and results throughout the life sciences are disseminated. Providing access to the biomaterials that form the basis of discoveries is critical, both to lower the cost of follow-on research and to identify potential errors or ambiguities associated with initial findings. Such independent verification of results is central to the scientific process. Many (though certainly not all) journals nominally require published authors to offer access to their materials, but very few of them actively enforce the requirement. A more proactive effort is needed. At a minimum, journal editors and scientific associations should develop electronic data archives and encourage the transfer of materials to BRCs. While the ultimate decision to deposit materials with a BRC should remain with the individual researcher, journals and scientific associations can and should try to increase researchers' incentives to make deposits by implementing and strictly enforcing access rules and by treating deposit as a preferred form of compliance.

—*In tandem with the federal government, the private sector must take a more proactive and direct role in supporting the life sciences research infrastructure.* U.S. international competitiveness in life sciences industries leverages an impressive record of public sector investment, which has built an active and complementary partnership between public and private sector research.[7] Though historically the private sector has supported BRCs through payments by individual companies for user fees and value-added services (for example,

7. Cockburn and Henderson (2001).

serving as international patent depositories), collective (but not anticompetitive) investments in areas of shared interest are crucial for long-term competitiveness. Industry associations such as PhRMA and BIO should invest in enhancing infrastructure, including the size, scope, and quality of BRC collections. Particularly for patent-protected research tools, industry associations should also fund BRC buyouts of key materials and their IP rights, thereby enabling the BRCs to lower license fees for all firms, enhancing research productivity and promoting effective cumulative research across firms.

—*Both university and industry associations should promote the use and flexibility of materials transfer agreements for deposits required by international patent regulations.* In addition to prohibiting unauthorized imitation, patenting creates a property right that can be bought and sold in the market. Standardized materials transfer agreements provide a cost-effective, efficient means for offering nonexclusive licenses to use patented materials, particularly those that were developed through university research or that are used primarily in basic research. Though in many cases maintaining exclusive control is preferred for commercial purposes, reducing the cost of a license enhances the potential for earlier distribution of materials for follow-on research.

Authentication Policies

The value of experimental research ultimately depends on acceptance of its findings by scientists other than those responsible for the research. Even when a claim is true, its impact on subsequent scientific research depends on the confidence that the scientific community has in it. Relative to other scientific disciplines, the life sciences are experimental science, and the body of knowledge at any point depends on the accumulation of evidence from experiments over time. It is nearly impossible to determine the merit or importance of a discovery exclusively through "thought experiments"; instead, analytical insight must be complemented by empirical evidence. Even with commercially motivated research, acceptance of the validity of findings by others is a prerequisite for licensing, establishing a strategic alliance (perhaps with a larger company), or receiving regulatory approval to market drugs, devices, or other technologies.

From the perspective of an individual researcher or company, the need to gain the acceptance of other experts in the field provides a strong incentive to use careful research methods and transparent experimental designs. However, the impact of faulty experimental methods or designs can extend far beyond an individual researcher's status or reputation. A single scientific research error—such as those caused by the use and distribution of unauthenticated biological materials—casts doubt on all research findings employing similar methods or materials. An error in a single experiment lowers the confidence of the scientific community in the research behind it and any subsequent research that builds on it. Moreover, as in the HeLa scandals, career and commercial concerns may limit the incentives of a researcher to admit an error once a finding is published. Finally, the costs of scientific error are enormous. A lower level of confidence in prior research findings necessitates the costly replication of past and current research before further progress can be made.

From the perspective of a single researcher, neither the scientific nor the commercial sector places sufficient value on the confirmation of prior results. However, the entire research community faces higher costs when research is not validated and authenticated. The scientific and commercial research communities can directly contribute to increasing the level of authentication by encouraging the use of standardized reference materials and encouraging the development of certification tools and techniques.

—The scientific community should adopt measures to reward techniques that facilitate rapid assessment of specific scientific claims, authenticate novel materials, and detect potential contamination or misidentification of materials. Scientific societies, journal editors, university departments, and industry associations should provide explicit policy and monetary support for developing tools and techniques within BRCs and related institutions. Further, the rapid disclosure of errors when they are discovered should be explicitly encouraged. For example, scientific journals should offer a brief editorial commentary on the speed and openness of researchers who offer retractions or corrections. Research that promotes the continued distribution of nonauthenticated materials or data that result in cumulative error should be strongly censured.

—Industry associations in the life and agricultural sciences must also proactively support the authentication and certification capabilities of BRCs and related institutions. In developing commercial applications of scientific research, the private sector depends heavily on the validity of published research. In most cases, the identification and development of science-based products relies on the authenticity of multiple claims throughout the scientific literature. For example, the development of combination AIDS therapies in the mid-1990s relied on synthesizing literally thousands of separate research findings across hundreds of scientific institutions. Since no individual researcher or company can authenticate all such findings, commercialization relies critically on the certification of central institutions, such as BRCs. Direct support of the development of technologies for certification and authentication enhances overall research productivity and promotes commercial opportunities.

Access Policies

Individual researchers choose materials (and suppliers) on the basis of their relative cost, efficiency, and quality. Researchers may not take into account the contribution of alternative material suppliers to the authentication and preservation of biological materials in general. Consequently, the ability of BRCs to serve as knowledge hubs depends on providing researchers with access to high-quality materials at a lower cost than that of peer-to-peer networks or for-profit distributors. The likelihood and terms of access to research materials, especially for materials protected by patent, depend on the norms and rules under which research is conducted. Both the scientific and commercial research communities have substantial influence in determining those rules and norms. These stakeholders should promote cumulative research by increasing access to research resources, for example, by implementing specific initiatives such as the following:

—*The scientific and commercial research communities should explicitly encourage independent and nondiscriminatory access to authenticated biomaterials and promote the use of standardized reference materials in experimental research.* In addition to encouraging the authentication of biomaterials and their deposit in a BRC, the scientific community must develop and actively promote the use of standardized materials in research. Among other things, the importance of biomaterials fidelity should be emphasized in training researchers on standard laboratory practice, and leading scientific organizations must work with policymakers and BRCs to establish international standards for authentication and preservation.

—*Scientific journals should require published authors to make research materials accessible on an independent and nondiscriminatory basis.* Journals should encourage the deposit of biomaterials in BRCs. In the absence of BRC deposit, journal editors should actively monitor whether researchers are making materials available for review and follow-on experiments. Besides physical materials, researchers should provide documented databases and other research resources (subject to appropriate confidentiality requirements). Journals should emphasize that accessibility augments the impact of a given research contribution (as measured by citations) and encourage the adoption of norms to enhance research method transparency and access to resources.

—*Universities and industry associations should explicitly encourage the use of MTAs as a means of providing access to materials while protecting intellectual property rights.* The increased use of patented materials throughout life sciences research both increases the amount of research conducted and potentially limits the availability of that research. Putting MTAs in place at the time of patent approval lowers the cost of mutually beneficial transactions between the developers of materials and follow-on researchers and widens the availability of patented biomaterials.

The scientific and commercial research communities have a direct stake in maintaining the progress of cumulative research. Scientific journals, scientific societies, and industry associations must not allow policy to be determined by individuals' desire to control, limit verification of, and restrict access to materials; instead, these stakeholders should advocate making the widest range of authenticated materials available to the broader research community.

Implications for Biological Resource Centers

Three interrelated factors drive the long-term effectiveness of BRCs as knowledge hubs for the life sciences: quality, operational efficiency, and reputation. First, and perhaps most important, the essential value of BRCs lies in the quality of the materials they provide, including the scale and scope of their collections, the precision of their authentication procedures, and the degree of access that they allow. BRC operational efficiency goes hand in hand with quality. Ineffective cataloguing or a slow distribution system, for example, reduces the likelihood that researchers will choose to obtain materials from a BRC, decreasing its potential impact. Finally, BRCs can enhance the productivity of the research community only if the research community has confidence in BRCs themselves. Without a strong reputation, the benefits of BRCs—including the reduced cost of duplication and increased use of standardized reference materials—will not materialize. BRCs are knowledge hubs, not simply biological product suppliers; as such, they must invest in activities that give the research community the highest level of confidence in their services and materials. Simply put, BRC operations should be guided by a "cumulative knowledge performance principle": to achieve their full potential as knowledge hubs for the life sciences, BRCs must build a strong record of quality and operational efficiency. Enhancing BRCs' quality, efficiency, and reputation has a multiplier effect, reducing costs and enhancing productivity across the research community.

As centralized facilities for materials exchange, BRCs can substantially enhance the fidelity of research materials used throughout the life sciences. However, the potential costs of an error on the part of a BRC are many times higher than the costs associated with an error by an individual researcher. Contamination or misidentification within BRC collections affects all researchers who draw on BRC materials, including later generations attempting to build on research based on BRC materials. Therefore, the research community depends on BRCs to be a "last line of defense" against the distribution of contaminated or misidentified biomaterials, but there are no systematic measures currently in place to monitor BRCs themselves.

Acquisition Policies

BRCs can achieve their potential as knowledge hubs for the life sciences only by focusing on acquisitions and activities that contribute directly to that role. However, BRCs may be subject to significant pressure to dedicate their limited resources to investments in noncore initiatives. For example, whereas the core purpose of a BRC is to serve the research community by providing well-characterized, authenticated materials, the Convention on Biological Diversity calls on national BRCs to implement a biodiversity initiative by maintaining large-scale repositories for the indigenous organisms of individual countries. Such initiatives would redirect BRCs' focus from making precise characterizations of a modest number of biomaterials for global distribution to dramatically expanding the comprehensiveness and scope of their collections. If BRCs take on significant responsibility for a biodiversity initiative, it is imperative that they do so without sacrificing quality standards throughout their collections.

—*BRCs should prioritize activities and investments according to their potential contribution to their central mission as knowledge hubs.* BRCs must avoid involvement in peripheral activities that might strain their ability to maintain quality and thereby reduce their reputation. BRCs should refuse to participate in noncore activities unless adequate funding is provided through public agencies or private associations. BRCs should grow slowly, carefully choosing new projects and acquisitions to establish an ever-higher level of quality and operational efficiency.

—*BRCs must actively help to establish and support high-quality BRCs around the world, especially in developing nations.* With the globalization and commercialization of life sciences research, the need for high-quality biological materials around the world is increasing. The establishment of regional BRCs around the world is central to developing the infrastructure required to meet that need. An international network of BRCs will complement the work currently performed by national BRCs, enhancing the quality of authenticated materials, increasing the degree of standardization of procedures and approaches to common problems, and expanding materials sharing across borders and organizations.

Authentication Policies

Although scientific journals and societies and individual researchers play an important role in exposing contamination and misidentification of materials, the unique position and capabilities of BRCs allow them to take the lead in authenticating deposits and identifying contaminated materials. Though the importance of authentication and standardized reference materials is acknowledged, BRCs face several challenges in establishing a systematic approach to authentication. First, to define appropriate protocols, BRCs must draw on current knowledge of key threats and identify novel problems and challenges as they arise. Close coordination with the research community itself is required, as early identification of contamination or misidentification depends on recognizing problems in experimental practice. In addition, BRCs must continually expand the range of the tools and technology that they use to authenticate materials. Advancements in information technology, along with greater understanding of biological systems, now allow for increasingly precise tests of biomaterials integrity. While enhancing the tools available for authentication, new technologies also raise their own challenges. For example, PCR has both dramatically enhanced the ability to identify materials through genetic screening and ushered in new threats, including the potential for viral contamination.

Ultimately, BRCs can enhance research productivity only if authenticated materials are adopted by the broader research community. BRCs must therefore take the lead over the long term in educating researchers on the problems that can arise in using biological materials. Ironically, the very success of BRCs in raising the fidelity of biomaterials raises the possibility that the research community itself will exercise less caution in using them. BRCs must guard against this possibility by adopting an educational mission to emphasize the risk of biomaterials contamination and misidentification in research practice.

Overall, BRCs can undertake several specific initiatives to reinforce their authentication function and to enhance their role as knowledge hubs for the life sciences:

—*The BRC community must focus on enhancing the overall availability and quality of authenticated materials throughout the life sciences.* While each of BRCs' core activities involves substantial costs, those costs are modest compared with the costs of the duplication, error, and lower productivity that result from the misuse of biological materials in published (or patented) research. To ensure quality and researchers' confidence in their collections, BRCs themselves must adopt advanced technologies to facilitate authentication and clarify the procedures they use to verify the integrity of their collections. When considering investment or participation in new activities, BRCs must be careful to avoid expansion into activities that might compromise the quality of authenticated materials—or their reputation as "honest brokers" in the broader scientific community.

—BRCs must invest in research to improve authentication and certification procedures, including the use of bioinformatics to increase the integration of physical collections and digital biological databases. Knowledge cannot be built upon unless it is verified and stored. BRCs and the research community have made dramatic strides over the past fifty years in developing technologies and protocols for each of BRCs' core functions, and investments in advancing certification technology must continue to be a high priority. It is crucial that BRCs provide clear guidance to the research community about the materials they supply. For example, though BRCs do inform users that certain deposits have "HeLa characteristics," leading BRCs should step up their campaign to educate researchers about the dangers of contamination and misidentification and aim to eradicate the inappropriate use of falsely labeled biological materials.[8] In this effort, the link between materials certification and bioinformatics will become increasingly important. Effective exploitation of bioinformatics may well require sophisticated knowledge management capabilities. BRCs are in a good position to lead the research community in developing protocols and new institutions to allow follow-on researchers to draw confidently on digital knowledge and to ensure that appropriate credit is given to those whose discoveries are being exploited.

—International BRCs must cooperate in establishing common standards and protocols for identifying and assessing authenticated biological materials. The global life sciences community benefits from the development of international standards for biological materials. Concordance between differing *national* standards should be a principal medium-term objective of the emerging international BRC network. While BRCs do occasionally compete for individual collections or for particular contracts, they mostly reinforce each other's integrity and reputation. BRCs should enhance their level of mutual cooperation and continue to invest collectively in areas of shared interest.

—Leading BRCs should "export" their expertise and capabilities in helping to establish international quality standards in BRCs throughout the world, particularly in newer BRCs in developing nations. More experienced BRCs should provide leadership, training, and technological assistance as new BRCs are established around the world, particularly with respect to national collections being

---

8. BRCs must be proactive in addressing the concerns of some biologists that the continued distribution of materials with "HeLa characteristics" is to blame for the continued inappropriate use of these materials (Masters, 2002). While our analysis suggests that BRCs have historically provided a better foundation for cumulative advancement and certification than the peer-to-peer network—and there are appropriate reasons at least to store cell lines that have been contaminated (for example, to undertake validation studies)—enhancing warnings about the use of cell lines with HeLa characteristics may enhance confidence in the materials throughout a BRC collection.

established under the Convention on Biological Diversity. Without effective training, investments in new culture collections will be ineffective, or, worse, reduce the integrity of and confidence in materials used in experimental research. By providing upfront assistance to new national collections, BRCs can establish and reinforce international quality standards, diffuse knowledge on efficient and effective laboratory techniques, and promote the adoption of standardized identification and certification systems (see box 5-4).

Access Policies

The value of authenticated BRC collections cannot be realized in the absence of long-term access. An access system must accomplish a number of related objectives, including the long-term preservation of authenticated materials, the distribution of materials on an independent and nondiscriminatory basis to qualified researchers, the legal transfer of materials protected by intellectual property rights, and compliance with transfer regulations and reporting requirements. For most BRCs, modest user fees play an important role in recovering the costs of distribution; in some collections, particularly in the United States, user fees also cover the costs of acquisition and authentication.

BRCs face both operational and strategic challenges in developing appropriate access policies. BRCs can achieve their potential only if they compete effectively in the market for biomaterials. As emphasized earlier, most BRC user requests are for a small number of deposits, and most "popular" materials are also available competitively from for-profit culture distributors. Though they must maintain the highest quality standards, BRCs must also achieve a high level of operational efficiency and offer value-added services to academic and commercial users.

At the same time, BRCs must guard against taking advantage of opportunities that come at the expense of their long-term reputation. Because of their central position in life sciences research, BRCs have numerous chances to gain short-term revenues or windfalls. For example, once a BRC takes control of a unique collection, it could charge the monopoly access price rather than encourage maximal diffusion by pricing at marginal cost. Though profitable in the short term, exploitation comes at the expense of a BRC's credibility as an honest broker in life sciences materials exchange. As knowledge hubs for life sciences research, BRCs must be vigilant in protecting their long-term reputation.

—*BRCs must focus on offering an access system that emphasizes their unique capabilities and advantages, advances nonexclusive access to biomaterials, and avoids direct competition in activities better accomplished by for-profit companies.* To maintain their position as effective knowledge hubs, BRCs must

---

Box 5-4. MR4: Addressing the Global Costs of Malaria

Malaria, one of the most devastating of the world's epidemics, is a central health and social challenge for developing nations. Responsible for at least 1 million deaths a year, malaria is a leading cause of death among children worldwide. Along with AIDS and tuberculosis, malaria hinders the development of robust economic and political institutions and the prospects for economic growth. More than 40 percent of the world's population is at risk for malaria, with the overwhelming majority living in developing countries (World Health Organization, 2003).

Over the past several decades, the level of research on malaria has been extraordinarily low compared with that on conditions also affecting advanced economies, such as heart disease, cancer, and AIDS. In large part, the low level of spending results from the combination of inadequate demand (in terms of the ability to pay) and continuing uncertainty about intellectual property rights; however, recent policy initiatives have attempted to provide focused incentives to enhance research and development of treatments for malaria and other tropical diseases (Kremer, 2001; Lanjouw, 2003).

The development of effective treatments for malaria has historically been hampered by inadequate research infrastructure. While many projects may have been promising in isolation, malaria research has suffered from inadequate coordination across research efforts and inconsistent procedures and protocols in both basic research and clinical evaluation (Anderson, MacLean, and Davies, 1996).

Beginning in the mid-1990s, the Wellcome Trust, along with other philanthropies and government agencies participating in the Multilateral Initiative on Malaria (MIM), began to make systematic investments to enhance malaria research, ranging from basic research on parasites to large-scale clinical projects in developing countries. Participants within MIM recognized the crucial importance of authenticated biological reagents in research, and, with funding from the National Institute of Allergy and Infectious Diseases,

---

avoid entering markets in which revenue streams depend on exclusive access or novel services better provided by the private sector. For example, BRCs should be at the forefront in establishing nonproprietary and nonexclusive bioinformatics standards to facilitate precise links between physical materials and genomic databases. BRCs also must avoid entering new business areas that compete directly with emerging for-profit companies in the bioinformatics industry.

the Malaria Research and Reference Reagent Resource Center (MR4) opened in 1999 as a biological resource center devoted entirely to the study of malaria.

Maintained by the ATCC, MR4 has come to fill a critical need in the scientific community for global access to standardized and authenticated research materials relating to malaria. The center provides parasite, host, and vector reagents to research scientists at the marginal cost of distribution. All charges for the collection, except for shipping and handling, are waived. MR4 also undertakes authentication research on malarial specimens, promotes technology transfer between research centers, and provides specific training in the appropriate use and application of malaria research resources. Many of its operations directly reflect the standard practices established by ATCC. For example, a standardized registration system allows depositors to maintain intellectual property rights over materials while facilitating access by the broader research community.

The MR4 program explicitly encourages links between field-based and laboratory-based research programs, especially within Africa. The MR4 Program has taken a number of steps—from running advertisements to sponsoring workshops and public events—to enhance access to its collections and expertise in African economic development and life sciences research communities (Zicker, 2001).

Despite the substantial progress made by MR4 in establishing a collection of reagents for malaria research, it faces a funding challenge over the long run. Particularly with the development of programs to enhance demand for tropical disease treatments, both government and industry now have a stake in developing an effective malaria and tropical disease research infrastructure. By taking collective action to ensure funds for acquisition, authentication, and preservation of biomaterials, these stakeholders have a promising opportunity to enhance the production of cumulative knowledge that has real potential for global impact.

—User fees should be set to recover the costs of distribution only. Over the long term, fixed costs associated with acquisition, authentication, and preservation should be raised from alternative sources, including depository fees, government funding, and collective funding through industry associations. Pricing at or near the marginal cost of distribution lowers research costs directly (through lower prices) and encourages more rapid diffusion of authenticated biological materials. While the monetary costs of research materials are usually small

relative to the overall costs of the average research project, the costs of misidentification or error can be enormous, particularly when errors go undetected prior to publication or patent approval. Rather than relying on cross-subsidization to recover fixed costs, BRCs should seek to obtain or renew funding that directly contributes to acquisition, authentication, and preservation. As emphasized earlier, an expanded government budget for research and collective funding through industry associations is critical.

   —*MTAs can facilitate nonexclusive access to research resources protected by intellectual property rights. International patent depositories should actively encourage nonexclusive licensing, particularly for materials with applications as research tools.* As more research resources come under patent protection, enhancing the environment for nonexclusive licensing of materials has become more important. By developing a "menu" of standard MTA agreements, BRCs can greatly reduce the costs of executing such licenses, both for materials developers and potential licensees.

   —*The objectives of new biosecurity requirements can be best achieved if BRCs assume primary responsibility for the physical transfer of regulated agents.* Effective biosecurity depends on careful and detailed documentation of the distribution of selected agents. Monitoring and investigation of bioterrorism threats will be enhanced by exploiting a hub-and-spoke system for materials exchange. Though new regulations restrict ownership and distribution of some agents, the implementation of an effective enforcement and monitoring system has only just begun. BRCs can serve as key intermediaries, simultaneously offering advanced documentation services and lowering the cost to individual researchers of complying with regulations. BRCs can build on their position to enhance their operational effectiveness and use of information technology to track patterns of access to selected agents. Coordination between BRCs and regulatory agencies will reduce the potential for improper transfers and increase their ability to investigate breaches of protocol when they occur. By building on their capabilities as hubs for materials exchange, BRCs can and should play a central role in establishing a robust system for access to selected agents.

   Biological resource centers have emerged as key players in the production of cumulative knowledge within the life sciences. Policymakers, the research community, and BRCs themselves have an important stake in ensuring the health and vitality of BRCs in this process. Indeed, because the life sciences themselves are at an early stage of development, future generations of scientists depend critically on the research being conducted today.

6

Future Challenges

This volume has provided a systematic assessment of the impact of biological resource centers on the production of cumulative knowledge. The analysis has highlighted the crucial role played by BRCs and related institutions in fostering scientific advancement over the twentieth century through the acquisition, authentication, preservation, and distribution of biomaterials. Simply put, BRCs serve as cumulative knowledge hubs for life sciences research. Indeed, as demonstrated in chapter 4, BRCs are not mere repositories of information; they magnify the impact of published research. At current levels of spending, enhancing the stock of authenticated biomaterials is likely to have a high rate of return in terms of enhancing cumulative scientific progress.

This final chapter outlines the across-the-board recommendations of this analysis and offers some concluding thoughts about the more general policy agenda for ensuring the strength of institutions that enable cumulative knowledge production.

Cross-Cutting Recommendations

The productivity of life sciences research depends on investment and participation in supporting the biomaterials exchange system—BRCs in particular. Stakeholders in both the public and private sectors benefit from an enhanced environment for biomaterials exchange; moreover, public and private investments complement each other by reinforcing different aspects of BRC activities. Building a strong life sciences research infrastructure depends on having a strong program of sustained investment over time. Such a program can help BRCs meet the contemporary policy challenges discussed earlier in the volume: bioinformatics, biotechnology, bioterrorism, and biodiversity.

Bioinformatics: Information Technology in the Life Sciences

Knowledge cannot be built upon unless it is stored and authenticated. While most policy analysis relating to bioinformatics focuses on the substantial challenges in implementing new technologies, the institutions verifying and preserving digital knowledge are of crucial importance. The effective exploitation of bioinformatics will require not only sophisticated knowledge management capabilities but also the development and adoption of standards. BRCs should take a leadership role in developing protocols that allow for concordance between digital databases and physical biomaterials. At the same time, BRCs must be sure to maintain their reputation as honest brokers, restricting their activities to those that are not better performed by for-profit companies.

Biotechnology: Private Commerce or Public Science?

Though commercial research and public science may seem at odds with each other, BRCs serve as powerful influences in ameliorating these tensions through their role as international patent depositories. The increased use and standardization of material transfer agreements (MTAs) offers a concrete mechanism for increasing access to research materials protected by intellectual property rights and provides a balance between public science and private commercialization efforts. The patent depository function also gives BRCs themselves a more stable financial foundation, insulating their budgets from transient political or economic pressures. While the role and impact of BRCs in promoting effective protection of intellectual property rights is still evolving, the establishment of international patent depositories in BRCs may be considered among the important policy achievements of the last twenty years.

Biosecurity: The Increased Threat from Bioterrorism

BRCs should serve as the central mechanism to regulate the transfer and distribution of materials that present biosecurity concerns. Policy analysis must move beyond the false trade-off between promoting science or ensuring national security. As a practical matter, most material exchanges in the life sciences (even of biohazardous materials) have been relatively haphazard and unsystematic. A structured biomaterials exchange system enhances both the prospects for cumulative research and national security. In other words, the knowledge required for an effective bioterrorism policy for biohazardous materials is the very knowledge required by a research community to conduct cumulative research.

Biodiversity: Preserving Nature

Biodiversity initiatives offer a key challenge to BRCs, in large part because their scale and scope far exceed BRCs' historical investments in their collections. The enormous requirements of an effective biodiversity program are beyond the capabilities and capacity of any single BRC; such a program requires the sharing of responsibilities through the development of an international BRC network. In order for a global network of BRCs to successfully undertake biodiversity initiatives, long-term funding (not isolated grant support) is a prerequisite. Moreover, establishing a global network of BRCs raises much more than a simple question of financing. An effective BRC network needs to ensure a level of expertise and appropriate capabilities at each node. Biodiversity efforts require substantial and sustained funding along with the "export" of governance and knowledge management skills of successful BRCs. Successful implementation of a multilateral effort holds the only hope that BRCs can expand their collections in a manner commensurate with biodiversity efforts while simultaneously ensuring the quality of their collections.

Concluding Thoughts

This volume has considered the impact of a specific type of institution—biological resource centers—on the environment for cumulative research in the life sciences. Though BRCs are important in their own right, they also provide a concrete example of a more general (and understudied) phenomenon: the role of economic institutions in shaping cumulative scientific and technical progress.

Two central insights expounded in this volume shed new light on the importance of economic institutions in facilitating the cumulative process. First, cumulative research depends on access to materials that allow follow-on researchers to validate and build on previous findings. Individual discoveries do not by themselves contribute to cumulativeness; the impact of research depends instead on future researchers being both aware of results and confident in their accuracy. In many (if not most) cases, these prerequisites cannot be satisfied unless the materials underlying research findings are made available to prospective follow-on researchers. This perspective on the conditions required for cumulative knowledge contrasts sharply with most economic and policy analysis, which simply assumes that cumulative knowledge will be produced. By the very nature of the scientific process, knowledge cannot be made "abstract" unless there are concrete manifestations by which to assess the validity of the knowledge.

Second, the challenges in establishing the validity of research findings are economic rather than technological or scientific. Though the scientific community often treats validation problems as if they can be "solved" by improvements in technology, the fundamental source of inadequate attention to validation is the incentive system within which scientific and commercial research is conducted. Individual researchers simultaneously face high-powered incentives to test the limits of technology in order to establish new research findings and too few incentives to ensure the validity of those findings. While the substantial rewards for new findings make the process of discovery extremely competitive, the confirmation of previous findings is mostly a public good. In other words, without changes in the incentives provided to scientific researchers, improvements in validation technology may have little impact on the rate of scientific error.

As a result, economic institutions play a central role in establishing the validity of research findings and enabling cumulative progress. While incentives in research are based overwhelmingly on commercial and scientific priority, cumulative knowledge production depends on making authenticated research findings available to the widest possible set of researchers. While numerous ways to organize the research authentication process are feasible, a centralized institution offers enormous potential benefits. Rather than having each researcher (or multiple groups of researchers) conduct individual validation studies (with little concordance among them), authentication through centralized institutions reduces the costs of achieving a given level of confidence. In other words, it is possible to provide institutions with incentives for authentication that cannot be replicated through an incentive system for individual researchers. Consequently, establishing and supporting institutions that provide nondiscriminatory and independent access to materials required to authenticate research helps to overcome the externalities associated with the scientific validation process.

It is important to emphasize that while this book has focused on the case of biological materials, the role of authenticated research materials and the scope of knowledge hubs is likely broader. From shared research resources such as linear accelerators in physics to genomic databases in biology to data sets used across the social sciences, scientific progress depends on the ability of researchers to access the materials underlying research findings, and, in so doing, ensure the integrity of the research process itself.

Accordingly, while attention in this volume has been confined to the specific challenges and opportunities associated with biological resource centers (and related institutions for biomaterials exchange), the policy argument has implications for science and technology policy more generally. Public policy toward science and innovation should be based on the ability of interventions to enable cumulative knowledge production and in so doing to enhance the productivity of multiple generations of researchers. While numerous programs and policies do support the cumulative process, inadequate attention has been paid to the distinction between expenditures on research per se and investments in cumulative knowledge production. While the development of a comprehensive policy and agenda designed to produce cumulative knowledge is beyond the scope of this book, the lessons learned from biological resource centers suggest that further research in this area holds a great deal of promise.

Appendix A: Econometric Results

A s described in box 4-3 and Furman and Stern, the data for the "marginal" analysis consist of 127 sets of BRC-linked special collections articles along with two sets of matched control articles.[1] The data include citation data as well as article, deposit, and author characteristics. Because citation data are realized in the form of annual "count" data and are highly skewed to the right (that is, the median is substantially smaller than the mean), the use of a traditional linear regression model (such as OLS) is inappropriate. Estimates will be downward biased as the analysis will "overweight" the high prevalence of observations for which the number of annual citations is equal to zero. While a Poisson estimation is the most traditional approach for dealing with count data, research over the past two decades suggests that the strong restrictions of the Poisson model (specifically, that the mean and variance of the underlying count data distribution are equal) can yield misleading results.[2] Instead, an appropriate and commonly used specification is the negative binomial regression model (a Poisson model in which the variance differs from the mean). For a

1. Furman and Stern (2003).
2. Wang, Cockburn, and Puterman (1998).

complete discussion of the advantages of the negative binominal in this context, see Cameron and Triverdi as well as Furman and Stern.[3]

Table A-1 presents the core findings. For each column, the dependent variable is annual forward citations, and each specification includes year fixed effects, vintage fixed effects, and either article-pair or article fixed effects (Furman and Stern includes a complete set of specifications, some of which exclude the control variables).[4] In addition to the variables defined earlier in the text, we define a new variable, "BRC-linked, post-deposit," which is equal only to 1 for those article-years after the materials associated with the article have been deposited at the ATCC. Regressions (1) and (2) include both BRC-linked and BRC-linked, post-deposit, allowing us to separate out the selection effect (that is, the BRC-linked variable captures the degree to which articles have higher citation rates independent of whether they have been deposited) from the marginal impact of BRC deposit (captured by the coefficient on BRC-linked, post-deposit).

Regression (1) presents an ordinary least squares (OLS) specification of forward citations on the BRC-linked and BRC-linked, post-deposit variables with control variables included. The results are significant: According to the OLS specification, articles associated with BRC-deposited materials received 57.7 percent more annual citations; after those materials were accessioned into ATCC, associated articles received a 46.5 percent boost in citations. This suggests that important articles are "selected" into BRCs and that BRCs have an important impact in enhancing the diffusion of knowledge associated with these articles.

As discussed earlier, however, OLS is inappropriate for count data. We therefore turned to negative binomial specifications. For these regressions, coefficients are reported as incidence-rate ratios (that is, a coefficient equal to 1 implies no effect while a 1.50 coefficient implies a 50 percent boost). The results in regression (2) confirm with a negative binomial specification what (1) suggests with OLS. Both the selection effect and the marginal impact of BRC-deposit are statistically and economically significant—the findings show that BRC-linked articles are twice as likely to be cited as control articles.

The final two columns report the core findings associated with the difference-in-differences analysis. In these models, we omit the BRC-linked dummy and employ article fixed effects (rather than article-pair fixed effects) to control for article-specific heterogeneity. The coefficient BRC-linked, post-deposit, again reflects the post-deposit citation boost experienced by

---

3. Cameron and Triverdi (1998) and Furman and Stern (2003).
4. Furman and Stern (2003).

Table A-1. *Summary Statistics*[a]

Variable	Mean	Standard deviation	Min.	Max.
Article-year measures (*N* = 13,947)				
Forward citations	5.77	12.65	0	186
BRC-linked	0.38	0.48	0	1
Age	9.10	6.56	0	31
Article deposition and author characteristics (*N* = 752 to 844 articles)				
Number of pages	6.69	5.59	0	70
Number of authors	4.86	3.69	0	57
University	0.59	0.49	0	1
Government	0.13	0.34	0	1
Foreign	0.34	0.48	0	1
Article characteristics[b] (*N* = 127)				
ATCC price (dollars)	229.80	44.01	167	270

a. Citation data were gathered on an annual basis from the year of publication through the end of 2001. The date of initial publication varies from 1970 to 1999, and the date of accession (deposit) into ATCC varies from 1981 to 1999.

b. Deposit year and price data (in dollars) exist only for articles associated with deposits to the ATCC special collections. All other data are from the base sample.

BRC-linked articles. In these models, the accession of materials into ATCC also was demonstrated to have an important impact on associated articles. Controlling for all time-invariant article-specific characteristics, ATCC deposit is associated with an 82 percent boost in citations. Expanding on the results in the earlier analysis, regression (4) demonstrates that the impact of deposit also increases over time. Overall, the econometric evidence provides support for a model in which biological resource centers play a key role in the accumulation of knowledge in the life sciences.

Table A-2. *Regression Results*[a]

Dependent variable = forward citations

	Ordinary least squares[b] (1) Overall BRC effect	Negative binomial		
		(2) Baseline count model	(3) Core difference-in-differences model	(4) Interaction between ATCC article * time since deposit
Article characteristic				
BRC-linked	0.577 (0.050)	2.204 (0.303)		
BRC-linked, post-deposit	0.465 (0.054)	2.023 (0.277)	1.824 (0.199)	1.677 (0.166)
BRC-linked * time since deposit				1.025 (0.012)
Control variable: Parametric restrictions[c]				
Article pair FEs = 0	126 χ^2, 4.7*10^5 p value, 0.00	65 χ^2, 3.6*10^{11} p value, 0.00		
Article FEs = 0			63 χ^2, 1.4*10^7 p value, 0.00	64 χ^2, 6.8*10^6 p value, 0.00
Age FEs = 0	30 χ^2, 20.98 p value, 0.00	30 χ^2, 265.66 p value, 0.00	30 χ^2, 417.85 p value, 0.00	30 χ^2, 418.83 p value, 0.00
Year FEs = 0[d]	23 χ^2, 2.24 p value, 0.00	23 χ^2, 58.94 p value, 0.00	23 χ^2, 67.95 p value, 0.00	23 χ^2, 67.37 p value, 0.00
Regression statistics				
R^2	0.52			
Log likelihood		−16,246.01	−14,396.54	−14,385.79
Number of observations	7,198	7,198	7,198	7,198

a. Robust standard errors are in parentheses. Coefficients reported as incidence-rate ratios; dependent variable = forward citations.

b. Dependent variable = ln(forward citations − 1).

c. The three values in each cell below correspond to the number of restrictions; χ^2; and p value.

d. Year FEs included for 1980–2001; 1970–74 and 1975–79 are grouped.

Appendix B: Biological Resource Centers: Policy Implications for Stakeholder Communities

Functional area	Funding agencies	Scientific and commercial research community	BRC community
Deposit	BRC deposit should be treated as de facto compliance with federal grant requirements for independent and nondiscriminatory access to biomaterials Funding should be provided for "buyouts" of high-priority acquisitions protected by IP	Journal policies should require authors to deposit biomaterials as de facto compliance after publication The private sector must take on a more direct role in supporting the life sciences research infrastructure University and industry associations should enhance the use of material transfer agreements (MTAs) for international patent depository materials	BRCs should focus on activities that advance their core mission as knowledge hubs The BRC community must assist in the development of BRCs around the world, especially those in developing nations

continued on next page

Functional area	Funding agencies	Scientific and commercial research community	BRC community
Certification/ authentication	The criteria for funding basic research grants should emphasize cumulativeness by including the use of and reliance on standardized reference materials The criteria for funding basic research grants should also include nondiscriminatory access to authenticated biomaterials, documented data sets, and other research resources Investments enhancing the fidelity of research resources should be substantially expanded over time	The scientific community must adopt reward systems that facilitate authenticating material claims and detecting contamination and misidentification Industry associations must take on a more proactive role in supporting certification capabilities of infrastructure institutions	The BRC community must focus on enhancing availability and quality of authenticated materials BRCs should continue to invest in research that improves their capabilities (such as bioinformatics) in authentication The international BRC community must cooperate in establishing common authentication protocols Leading BRCs should "export" their expertise and quality standards to BRCs throughout the world, particularly those in developing nations

continued on next page

Functional area	Funding agencies	Scientific and commercial research community	BRC community
Access	Federal funding must continue to require researchers to offer ongoing access to novel materials	The scientific and commercial research communities should encourage access to authenticated biomaterials and promote the use of standardized reference materials	BRCs must focus on offering nonexclusive access rather than engage in activities better accomplished by for-profit providers
	The capability to preserve biomaterials should be substantially expanded over time	Journals should require authors to make research material available on a nondiscriminatory basis	User fees should be set to cover the cost of distribution only, with fixed costs funded through alternative sources
	Effective access to biomaterials is facilitated by a competitive market with observable standards for quality and safety	University and industry associations should encourage use of MTAs to provide access while protecting IP	BRCs that are also international patent depositories should encourage wider access facilitated by MTAs
	USPTO should facilitate nonexclusive and independent licensing of materials held in international patent depositories		BRCs must assume primary responsibility for the transfer of select agents with biosecurity concerns
	BRCs should play a central role in ensuring biosecurity in the distribution of select materials		

References

Adams, James. 1990. "Fundamental Stocks of Knowledge and Productivity Growth." *Journal of Political Economy* 98 (4): 673–702.

Adams, James, and Zvi Griliches. 1996. "Measuring Science: An Exploration." *Proceedings of the National Academy of Sciences* 93 (23): 12664–70.

American Association for the Advancement of Science. 2002a. "AAAS Analysis Considers Impact of New Agency on Federal Funding for Science and Technology" (www.aaas.org/news/releases/2002/1126rd. shtml).

————. 2002b. "An Overview of Federal Funding of Biodefense Research for FY 2003" (www.aaas.org/spp/rd//pr120302.pdf).

Anderson, J., M. MacLean, and C. Davies. 1996. *Malaria Research: An Audit of International Activity*, PRISM Report No. 7. Wellcome Trust, London.

Angrist, Joshua D., Guido W. Imbens, and Donald B. Rubin. 1996. "Identification of Causal Effects Using Instrumental Variables." *Journal of the American Statistical Association* 91 (434): 444–72.

Atlas, Ronald. 1999. "The Threat of Bioterrorism in America: Assessing the Adequacy of the Federal Law Relating to Dangerous Biological Agents." Testimony before the Committee on Commerce, U.S. House of Representatives, Washington, May 20.

Audretsch, David, and Paul Stephan. 1996. "Company-Scientist Locational Links: The Case of Biotechnology." *American Economic Review* 86 (June): 641–52.

Baker, Nicholson. 1996. "Discards," *New Yorker*, October 14, pp. 64–86.

Barabasi, Albert-Laszlo. 2002. *Linked: The New Science of Networks*. Cambridge, Mass.: Perseus Books Group.

Beck, David. 2001. "Building for the Huntington Disease Gene Discoveries." *Cell Collections* (Coriell Institute for Medical Research): 8–10.

Blaine, Lois. 2003. "Knowledge Management and Bioinformatics." In *Biological Resource Centers: Their Impact on the Scientific Community and the Global Economy*, edited by Raymond H. Cypess, pp. 17–24. Manassas, Va.: American Type Culture Collection.

Bush, Vannevar. 1945. *Science: The Endless Frontier*. Government Printing Office.

California State University. 2003. "Report of the Task Force on Library Collections" (www.calstate.edu/LS/Lib_Coll_Rpt.pdf [December 2003]).

Cameron, A. C., and P. K. Triverdi. 1998. *Regression Analysis of Count Data*. Cambridge University Press.

Candiotti, Susan, and Bill Mears. 2001. "FBI Profiler: We Will Catch Anthrax Mailer." CNN News Network, November 16, 2001 (www.cnn.com/2001/US/11/16/anthrax.profiler/).

Clark, William A., and Dorothy H. Geary. 1974. "The Story of the American Type Culture Collection—Its History and Development (1899–1973)." *Advances in Applied Microbiology* 17: 295–309.

Cockburn, Iain, and Rebecca Henderson. 2001. "Publicly Funded Science and the Productivity of the Pharmaceutical Industry." In *Innovation Policy and the Economy*, vol. 1, edited by A. Jaffe, J. Lerner, and Scott Stern. MIT Press.

Crewdson, John. 1995. "In Gallo Case, Truth Termed as a Casualty." *Chicago Tribune*, January 1, 1995, p. 1.

Cusumano, Michael, and Richard Selby. 1997. *Microsoft Secrets*. Free Press.

Cypess, Raymond H. 2003a. "Biological Resource Centers: Essential to Progress." In *Biological Resource Centers: Their Impact on the Scientific Community and Global Economy*, edited by Raymond H. Cypess, pp. ix–xi. Manassas, Va.: American Type Culture Collection.

———. 2003b. "The Vision." In *Biological Resource Centers: Their Impact on the Scientific Community and Global Economy*, edited by Raymond H. Cypess, pp. 1–5. Manassas, Va.: American Type Culture Collection.

———. 2003c. "Biosharing and Intellectual Property." In *Biological Resource Centers: Their Impact on the Scientific Community and the Global Economy*, edited by Raymond H. Cypess, pp. 36–42. Manassas, Va.: American Type Culture Collection.

Cypess, Raymond H., ed. 2003d. *Biological Resource Centers: Their Impact on the Scientific Community and the Global Economy*. Manassas, Va.: American Type Culture Collection.

Dasgupta, Partha, and Paul David. 1994. "Towards a New Economics of Science." *Research Policy* 23 (5): 487–521.

Environmental Health and Radiation Safety. 2001. "Complying with the U.S. Patriot Act of 2001" (www.ehrs.upenn.edu/protocols/patriot.html).

Ernst and Young. 2000. *Convergence: The Biotechnology Report* (www.ey.com).

Freshney, Ian R. 2000. *Culture of Animal Cells: A Manual of Basic Technique.* 4th ed. John Wiley.

Furman, Jeffrey S., and Scott Stern. 2004. "Climbing atop the Shoulders of Giants: The Impact of Institutions on Cumulative Research." Mimeo. Evanston, Ill.: Northwestern University.

Gans, Joshua, and Scott Stern. 2003a. "The Product Market and the Market for 'Ideas': Commercialization Strategies for Technology Entrepreneurs." *Research Policy* 32 (2): 333–50.

———. 2003b. *Assessing Australia's Innovative Capacity in the 21st Century.* Australian Institute for Commercialization.

Garrett-Jones, Sam. 2004. "From Citadels to Clusters: The Evolution of Regional Innovation Policies in Australia." *R&D Management* 34 (1): 3–16.

Gold, Michael. 1986. *A Conspiracy of Cells: One Woman's Immortal Legacy and the Medical Scandal It Caused.* State University of New York Press.

Griliches, Zvi. 1990. "Patent Statistics as Economic Indicators: A Survey." *Journal of Economic Literature* 92 (4): 630–53.

———.1998. *R&D and Productivity: The Econometric Evidence.* University of Chicago Press.

Griliches, Zvi, ed. 1984. *R&D, Patents and Productivity.* University of Chicago Press.

Grossman, Gene, and Elhanan Helpman. 1991. *Innovation and Growth in the World Economy.* Harvard University Press.

Hay, Robert J. 2003. "ATCC Historical Perspective." In *Biological Resource Centers: Their Impact on the Scientific Community and Global Economy*, edited by Raymond H. Cypess, pp. 153–61. Manassas, Va.: American Type Culture Collection.

Henderson, Rebecca, and Iain Cockburn. 2001. "Publicly Funded Science and the Productivity of the Pharmaceutical Industry." In *Innovation Policy and the Economy*, edited by A. Jaffe, J. Lerner, and S. Stern. MIT Press.

Holley, Edward G. 1987. "North American Efforts at Worldwide Acquisitions since 1945." *Collection Management* 9 (summer/fall): 89–111.

Ingram, Judith, and Sergei Shargodsky. 2002. "Russians Help Bioterror Defense," January 28, 2002 (www.cdi.org/russia/johnson/6045-8.cfm).

Jaffe, Adam, and Manuel Trajtenberg. 2002. *Patents, Citations and Innovations: A Window on the Knowledge Economy.* MIT Press.

Jaffe, Adam, Manuel Trajtenberg, and Rebecca Henderson. 1993. "Geographic Localization of Knowledge Spillovers as Evidenced by Patent Citations." *Quarterly Journal of Economics* 108 (3): 434, 577–98.

Jones, Charles. 1995. "R&D Based Models of Economic Growth." *Journal of Political Economy* 103 (4): 739–84.

———. 1998. *Introduction to Economic Growth.* Norton.

Jong, S. C. 2003. "Core Functions of Biological Resource Centers." In *Biological Resource Centers: Their Impact on the Scientific Community and the Global Economy*, edited by Raymond H. Cypess, pp. 6–16. Manassas, Va.: American Type Culture Collection.

Kaplan, Steven, and Mohanabir Sawhney. 2000. "E-Hubs: The New B2B Market-places." *Harvard Business Review* (May/June): 97–103.

Knobler, Stacey L., and others. 2002. *Biological Threats and Terrorism: Assessing the Science and Response Capabilities: Workshop Summary.* National Academy of Sciences.

Kremer, Michael. 2001. "Creating Markets for New Vaccines—Part I; Rationale." In *Innovation Policy and the Economy,* vol.1, edited by A. Jaffe, J. Lerner, and Scott Stern. MIT Press.

Lamoreaux, Naomi, and Kenneth L. Sokoloff. 2001. "Market Trade in Patents and the Rise of a Class of Specialized Inventors in the Nineteenth-Century United States." *American Economic Review* 91 (May, *Papers and Proceedings, 2001*): 39–44.

Lanjouw, Jean. 2003. "Intellectual Property and the Availability of Pharmaceuticals in Poor Countries." In *Innovation Policy and the Economy,* vol. 3, edited by A. Jaffe, J. Lerner, and Scott Stern. MIT Press.

Lee, Dan. 2003. "Sweden Forging Ahead on Stem Cells," March 30, 2003 (www.bayarea.com/mld/mercurynews/business/5517520.htm).

Mach, Marin A., and Mieczyslaw L. Owoc. 2001. "Validation as the Integral Part of a Knowledge Management Process." *Informing Science* (June 19–22): 346–51.

MacLeod, R. A. F., and others. 2001. "Widespread Intraspecies Cross-Contamination of Human Tumor Cell Lines." *International Journal of Cancer* 83: 555–63.

Marburger, John. 2002. "Comments by the President's Science Advisor, National Summit on the U.S. Scientific Enterprise: New Direction?" Council of Scientific Society Presidents. Washington. May 7.

Masters, John R. W., and others. 2001. "Short Tandem Repeat Profiling Provides an International Reference Standard for Human Cell Lines." *Proceedings of the National Academy of Sciences* 98 (14): 8012–17.

———. 2002. "HeLa Cells 50 Years On: The Good, the Bad and the Ugly." *Nature Reviews Cancer* 2 (4): 315–19.

Merton, Robert. 1973. *The Sociology of Science: Theoretical and Empirical Investigation.* University of Chicago Press.

Moel, Henk F., and others. 1985. "The Use of Bibliometric Data for the Measurement of University Research Performance." *Research Policy* 14 (3): 131–49.

Mokyr, Joel. 2002. *The Gifts of Athena.* Princeton University Press.

Murray, Fiona. 2002. "Innovation as Overlapping Scientific and Technological Trajectories: Exploring Tissue Engineering." *Research Policy,* vol. 31, nos. 8-9, pp. 1389–403.

National Academies. 2002. President's Address (www4.nationalacademies.org/nas/nashome.nsf/urllinks/NAS-59MUXF?OpenDocument).

National Institutes of Health. 2002. "Stem Cell Basics" (stemcells.nih.gov/infoCenter/stemCellBasics.asp).

National Science Board. 2002. *Science and Engineering Indicators 2002.* NSB-02-1. Arlington, Va.: National Science Foundation.

Nelson, Richard R. 1993. *National Innovation Systems: A Comparative Analysis.* Oxford University Press.

Nelson, Richard, and Nathan Rosenberg. 1994. "American Universities and Technical Advance in Industry." *Research Policy* 23 (3): 323–48.

Nuclear Threat Initiative. 2002. "Russia: Biological Resource Center Faces Bankruptcy," October 23, 2002. Global Security Newswire (www.nti.org/d_newswire/issues/thisweek/2002_10_23_biow.html).

Organization for Economic Cooperation and Development. 2001. *Biological Resource Centres: Underpinning the Future of Life Sciences and Biotechnology.* Paris: OECD Books.

Ostroff, Stephen. 1999. "The Threat of Bioterrorism in America: Assessing the Adequacy of the Federal Law Relating to Dangerous Biological Agents." Testimony before the Committee on Commerce, U.S. House of Representatives, Washington, May 20.

Owen-Smith, Jason, and Walter W. Powell. 2004. "Knowledge Networks as Channels and Conduits: The Effects of Spillovers in the Boston Biotechnology Community." *Organization Science* 15 (1): 5–21.

Powell, Walter, Kenneth Koput, and Laurel Smith-Doerr. 1996. "Interorganizational Collaboration and the Locus of Innovation: Networks of Learning in Biotechnology." *Administrative Science Quarterly* 41 (1): 116–45.

Powell, Walter W., and others. 2004. "Network Dynamics and Field Evolution: The Growth of Interorganizational Collaboration in the Life Sciences." *American Journal of Sociology,* forthcoming.

Rabinow, Paul 1997. *Making PCR: Story of Biotechnology.* University of Chicago Press.

Romer, Paul. 1990. "Endogenous Technological Change." *Journal of Political Economy* 98 (5): S71–S102.

Rosenberg, Nathan. 1963. "Technological Change in the Machine Tool Industry, 1840–1910." *Journal of Economic History* 23 (4): 414–43.

———. 1974. "Science, Invention, and Economic Growth." *Economic Journal* 84 (333): 90–108.

———. 1990. "Why Do Firms Do Basic Research (with Their Own Money)?" *Research Policy* 19 (2): 165–174.

Rosenkopf, Lori, and Michael L. Tushman. 1998. "The Coevolution of Community Networks and Technology: Lessons from the Flight Simulation Industry." *Industrial and Corporate Change* 7 (2):311–46.

Shapiro, Carl, and Hal Varian. 1999. *Information Rules.* Harvard Business School Press.

Shilts, Randy. 1993. *And the Band Played On: Politics, People and the AIDS Epidemic.* St. Martin's Press.

Simione, Frank P. 2003. "Regulatory Coordination and Compliance." In *Biological Resource Centers: Their Impact on the Scientific Community and the Global Economy,* edited by Raymond H. Cypess, pp. 30–31. Manassas, Va.: American Type Culture Collection.

Solow, Robert M. 1957. "Technical Change and the Aggregate Production Function." *Review of Economics and Statistics* 39 (6): 312–20.

Staley, James T. 2003. "Microbial Diversity and Biological Resource Centers." In *Biological Resource Centers: Their Impact on the Scientific Community and the Global Economy*, edited by Raymond H. Cypess, pp. 110–11. Manassas, Va.: American Type Culture Collection.

Stern, Scott. 2004 (forthcoming). "Do Scientists Pay to Be Scientists?" *Management Science*.

Stevenson, Robert E., and Harold Hatt. 1992. "Culture Collections, Functions." *Encyclopedia of Microbiology*, vol. 1, p. 15. San Diego: Academic Press.

Tyler, Patrick. 2002. "Russian Lab Storing Germs Faces Cutoff of Electricity." *New York Times*, April 7, p. A11.

USA Today. 2002. "Saudi Takes Lead on Stem-Cell Cloning" (www.usatoday.com/news/science/2002-07-09-arab-stemcell.htm).

U.S. House of Representatives, Subcommittee on Oversight and Investigations. 1995. *Gallo Report: Investigation of the Institutional Response to the HIV Blood Test Patent Dispute and Related Matters*. Government Printing Office. Available at www.aegis.com/law/journals/1995/gallo002.html (accessed May 2004).

U.S. House of Representatives. 2001. "Summary of the Public Health Security and Bioterrorism Response Act of 2001" (www.house.gov/commerce_democrats/bioterrorism/summary.htm).

Wall Street Journal. 2002. "Ex-Soviet Biological Weapons Lab in Demand." January 9, 2002.

Wang, P., I. Cockburn, and M. Puterman. 1998. "Analysis of Patent Data—A Mixed Poisson Regression Approach." *Journal of Business and Economic Statistics* 16 (1): 27–41.

Watson, James D. 1968. *The Double Helix: A Personal Account of the Discovery of the Structure of DNA*. New York: Atheneum.

Weiss, Rick. 2002. "Clues to Anthrax Attacks Found." *Washington Post*, May 10, p. A2.

World Health Organization. 2003. "Africa Malaria Report 2003" (mosquito.who.int/amd2003/amr2003/pdf/amr2003.pdf).

Zicker, Fabio. 2001. "Strengthening Capability for Malaria Research in Africa," *Emerging Infectious Diseases* 7 (3): 529–30 (www.cdc.gov/ncidod/eid/vol7no3_supp/zicker.htm).

Index